THE BIG IDEAS IN SCIENCE

A complete introduction

Jon Evans

First published in Great Britain in 2011 by Hodder Education.

This revised and updated edition published 2020 by Teach Yourself, an imprint of John Murray Press, a division of Hodder & Stoughton. An Hachette UK company.

British Library Cataloguing in Publication Data: a catalogue record for this title is available from the British Library.

Library of Congress Catalog Card Number: on file.

ISBN: 978 1 529 39795 6

Ebook ISBN: 978 1 529 34257 4

1

Typeset by Cenveo® Publisher Services.

Printed and bound in Great Britain by CPI Group (UK) Ltd, Croydon, CR0 4YY.

John Murray Press policy is to use papers that are natural, renewable and recyclable products and made from wood grown in sustainable forests. The logging and manufacturing processes are expected to conform to the environmental regulations of the country of origin.

Carmelite House

50 Victoria Embankment

London EC4Y 0DZ

www.hodder.co.uk

Contents

Introduction to the second edition

Nothing stays the same, and that is certainly true for both science and the title of this book. *The Big Ideas in Science* is actually the second edition of *Understand Science*, which was published under the Teach Yourself imprint in 2011. All 30 chapters have been fully updated to reflect the advances in science and technology that have occurred over the past decade or so.

Reflecting the vagaries of scientific discovery, some topics have seen more advances than others, so some chapters have changed more than others. However, every chapter contains a brand-new Spotlight box exploring a subject related to the main topic. There are new Spotlight boxes on subjects such as dark matter and energy, super-resolution microscopy, friendly bacteria, 3D printing and the dangers of solar flares. Each chapter also contains a list of key ideas summarizing the main take-home points and the details of at least two books that will allow readers to explore the topic of the chapter in more detail.

The one thing that has stayed the same is that I would again like to thank Sarah, Charlotte and Claudia for all their love and support.

Introducing *Homo scientificus*

In certain ways, we've hardly changed at all in the 200,000 years since modern humans (*Homo sapiens*) first strode manfully onto the scene. In essence, we look, behave and think much the same now as we did then. In other ways, though, we've changed beyond all recognition.

And these changes haven't been spread evenly over the past 200,000 years; they've mainly been concentrated into the past few hundred years. We may have developed agriculture, religion, culture, warfare and cities over those 200,000 years, but up until quite recently the life of the average person was much as it had always been. Then we developed science and change went into overdrive.

By the simple expedient of asking questions about the universe and then conducting experiments to answer those questions, science has completely transformed our world and our lives. It has revolutionized our understanding of the universe and our place within it, and allowed us to develop previously undreamt-of technologies.

We used to think that the whole universe revolved around us, both figuratively and in reality; we now know that Earth is just one of countless planets orbiting countless stars in countless galaxies. Even our universe may be just one among countless others.

Meanwhile, rather than being specifically designed, we now know that we (as every other organism on Earth) are simply the fortunate result of a blind, ratchet-like process that honed us for a specific environmental niche (which in our case was the African savannah). Still, that doesn't mean that we're not special; as far as we know, we could be the only conscious, sentient beings in the whole universe.

We have harnessed our growing scientific knowledge to develop generation upon generation of cutting-edge technologies. And these technological marvels don't just make our lives longer, easier and more enjoyable; they also confirm the accuracy of our scientific theories. We are able to build planes that fly reliably, design smartphones that communicate reliably and produce breakfast cereals that go snap, crackle and pop reliably because our scientific theories reflect fairly accurately how the universe operates.

Science has revealed a universe that is older, larger and stranger than we could ever have imagined. It has taken us back to the very dawn of time and forward to the end of all things. It has allowed us to probe the depths of reality and explore the vast expanses of space. It has shown us how to take control of the universe: how to fly, how to communicate over vast distances, how to lace our world in light.

Science has made us who we are and given us our modern world. We are fundamentally not the same creatures that strode out onto the African savannah 200,000 years ago. We are more than human; we are practically a new species: we are *Homo scientificus*.

This book is an introduction to *Homo scientificus* and what it has achieved over the past few hundred years. In the following 30 chapters, you will learn what science has discovered about matter, space, energy, life, weather, information and the future, and how we have transformed these discoveries into our modern technologies. You will witness the birth of the solar system, follow ocean currents for thousands of miles, ride on beams of light and find out why dolphins are perhaps the most perverted creatures on the planet.

But why should you bother with science at all when you're busy cooking eggs? You appreciate that science is important and you're grateful for all the technologies it has given you, but why do you need to know how it all works? The fact that it works at all is quite enough for you.

Well, it's precisely because science is so integral to our world – that we are now *Homo scientificus* – that you really need to grasp what science has revealed about the workings of the universe. You can't ignore science, because science doesn't ignore you.

For science directly affects our lives in so many ways, both good and bad. And to make informed decisions about where the balance between the good and bad aspects should lie, we need to understand science, its theories and the technologies that arise from it. In fact, many of the most important issues currently facing humankind have a strong scientific and technological bent.

Why is global warming happening and what can we do about it; is cloning an important new medical technology or an affront to humanity; does the development of artificial intelligence herald a great opportunity or a great threat, or both? If you want to have a say in how we deal with these challenges and advances, then you need to understand the science behind them.

And how we deal with these challenges and advances will determine the path that *Homo scientificus* follows into the future. Will it be long and glorious, or short and destructive? If science has achieved much in the past few hundred years, think how much more we could achieve over the next few hundred years, or few thousand. We could journey to the stars, live for centuries or enhance our minds and bodies in countless ways, creating a physical embodiment of *Homo scientificus*. Unless, of course, we destroy ourselves first.

But even if we manage to avoid destruction by our own hand, we are still ultimately at the mercy of the universe. Science has already revealed the many ends that we will face: the end of our solar system, the end of our galaxy, the end of our universe. Even the first of these endings – that of our solar system – is billions of years in the future, but to be forewarned is to be forearmed.

As *Homo sapiens* we would be compelled to follow the universe to its bitter end. But as *Homo scientificus*, perhaps we can find an escape route.

Part One

How we got here

1

Bang, we're off

Just under 14 billion years ago, there was nothing: no people, no planets, no stars, no space, no time. Nothing. Then, for reasons that are still unexplained, the entire universe suddenly and unexpectedly burst into being, in what scientists like to call 'the Big Bang'.

Initially, however, there was nothing really very big about it, because when the universe first popped into existence it did so within a volume that was much, much smaller than an atom. Almost immediately, however, the universe began to expand incredibly rapidly, in a process known as inflation, which then stopped almost as soon as it had begun. Despite lasting for just a minuscule fraction of a second, during this inflationary period the universe more than doubled in size 100 times, growing to around 30 cm.

The ending of this rapid period of expansion released a huge amount of energy, which had the handy consequence of creating all the matter that fills the universe today. But because this matter was squeezed into just 30 cm, the universe at this point was very different to the one we now inhabit.

Expanding and cooling

It consisted of an unimaginably dense soup of tiny particles at a temperature of 10^{27} degrees (which is a convenient way to write 1 followed by 27 zeros, or one thousand trillion trillion degrees). The universe didn't stay that way for long though; for although the inflationary period had ended, it kicked off a gentler period of expansion that continues to this day.

According to the latest astronomical measurements, the observable universe (meaning the universe we can see with telescopes) currently extends for around 46 billion light years (the distance that light travels in a year, or around 1 trillion kilometres) in all directions and is expanding at 73 km a second.

In fact, it was the discovery that the universe is expanding (see Spotlight) that led to the formation of the Big Bang theory for the origin of the universe. The argument goes that if the universe is expanding now then it must have been smaller in the past, meaning that if you trace the expansion back far enough then the universe essentially disappears up its own fundament, like playing film of an explosion backwards.

By the time the universe was one ten-thousandth of a second old, this expansion had caused the temperature to drop to one trillion degrees. While still unimaginably hot, this temperature was low enough for a number of important changes to start taking place in the now slightly less dense soup of tiny particles.

Spotlight: Edwin Hubble, 1889–1953

If any one scientist can be said to have ushered in the modern conception of the universe, both vast and getting vaster, it is the US astronomer Edwin Hubble.

Prior to his work in the 1920s, astronomers thought our Milky Way galaxy comprised the whole universe. By studying the light released by certain types of stars, Hubble first showed that some hitherto puzzling patches of stars were not in the Milky Way at all, but were actually individual galaxies millions of light years away. He then discovered that all these galaxies are moving away from us, with the furthest galaxies moving away the fastest, providing the smoking gun for the expanding universe.

Hubble was hardly a stereotypical scientist. A talented athlete in his youth, Hubble claimed that he once fought an exhibition boxing match against the French national champion. Then later, when his scientific discoveries had brought him a certain measure of renown, Hubble attended Hollywood parties with film stars of the day such as Charlie Chaplin and Greta Garbo.

His achievements are now commemorated in the form of the Hubble Space Telescope.

Quarks and leptons

By tiny particles, we mean subatomic particles: in other words, the particles that make up atoms. These particles, known collectively as fermions, can be divided into two broad groups: quarks and leptons. These two groups are further divided into three families, with the first family of each group containing the most common and familiar types of subatomic particle.

The most common quarks are known as up and down quarks, while the most common lepton is the electron. These most common subatomic particles form all the ordinary matter that we see around us – stars, planets, humans, cabbages. The less common subatomic particles (which include the muon, the tauon, neutrinos, and charm, strange, top and bottom quarks) make up more exotic forms of matter, which are generally only seen on Earth at the extremely high energies generated in particle accelerators (see Spotlight).

Spotlight: Particle accelerators

Piecing together the story of the beginning of the universe has involved a combination of astronomical observations, mathematical models and particle accelerators.

These giant and hugely expensive instruments probe the intricacies of the subatomic world by slamming particles such as electrons and protons into a static target or, more recently, into each other at very high speeds. The huge amounts of energy produced by these collisions create a wide range of other subatomic particles, including exotic particles such as neutrinos.

On 10 September 2008, the largest and most powerful particle accelerator yet built was officially switched on. Housed hundreds of feet beneath the ground near Geneva in Switzerland, the Large Hadron Collider (LHC) cost $10 billion to construct and consists of a ring 17 miles in circumference.

By sending protons around this ring in opposite directions and then smashing them together at almost the speed of light, scientists hope to create subatomic particles that have never been seen before. In 2012, these efforts famously resulted in the first detection of the Higgs boson, a particle thought to confer mass on the particles that make up ordinary matter.

The strong force

When the universe was younger than one ten-thousandth of a second, all these particles existed as independent entities in the extremely hot, dense soup. But as the universe expanded, the temperature dropped sufficiently for these particles to start joining together to form larger entities.

This is because temperature is a measure of the amount of energy present in a system and above a certain temperature the subatomic particles just possessed too much energy. They were colliding with such force that they simply rebounded. As the temperature dropped, however, so did the energy levels of the particles, allowing them to begin to stick together.

For quarks, this coming together was mediated by what is known as the strong force, which is one of the four fundamental forces of nature. The strong force started to operate when the temperature of the universe dropped below one trillion degrees, joining the up and down quarks together into collections of twos and threes.

As the temperature dropped further, the collections of two quarks fell apart, until all that were left were collections of two up quarks and one down quark and collections of two down quarks and one up quark. These collections now began to operate as particles in their own right, with the former collection of quarks becoming protons and the latter collection becoming neutrons. But this process ended up generating a great deal more protons than neutrons, because many of the neutrons decayed into protons.

After bringing together all the up and down quarks, the strong force then began to combine the protons and neutrons into various permutations. This included one proton and one neutron, two protons and one neutron, and two protons and two neutrons, with the latter combination being the most stable and long-lasting. Larger combinations quickly fell apart, except combinations containing three or four protons with a few neutrons thrown in, which were produced in small amounts.

All this occurred in the first three minutes of the universe's life, at which point its temperature had dropped to a balmy 1 billion degrees. Below this temperature, there was no longer enough energy around for the strong force to stick together any more protons and neutrons, but those that it had already stuck together remained together.

The universe then stayed like this for the next 380,000 years, steadily expanding and cooling. During this time, the universe consisted of various combinations of protons and neutrons, together with a load of excess single protons, and numerous electrons and other leptons all flying around ignoring each other.

The electromagnetic force

The main distinction between quarks and leptons is that quarks interact via the strong force, but leptons don't. Electrons, for instance, interact solely via the electromagnetic force. This is another fundamental force of nature, which switched on at about the same time as the strong force but was initially unable to have much of an influence because the temperature was too high.

The electromagnetic force acts between all particles that possess an electric charge. Such charged particles possess either a positive charge or a negative charge: particles with opposite charges attract each other, while those with identical charges repel each other.

It's like holding two magnets: opposite poles snap together, while identical poles force themselves apart.

Electrons are negatively charged and so are influenced by the electromagnetic force, as are protons, which are positively charged. Neutrons, on the other hand, do not have an electric charge and so are not influenced by the electromagnetic force.

By the time the universe was 380,000 years old, its temperature had dropped to 3,000 degrees and the electromagnetic force could start bringing together the oppositely charged protons and electrons to form atoms (together with any neutrons that accompanied the protons). All atoms comprise a nucleus of protons and neutrons surrounded by a cloud of orbiting electrons. And because atoms always possess an identical number of protons and electrons, they have no overall charge.

The first atomic matter

So, after 380,000 years, the first proper atomic matter finally appeared. Single electrons combined with the excess protons to form hydrogen, while pairs of electrons joined the combinations of two protons and two neutrons to form helium (see Figure 1.1). The small amount of larger combinations containing three or four protons also acquired electrons, producing lithium and beryllium respectively.

The end result of all this was the creation of a huge amount of hydrogen, accounting for around 77 per cent of the newly created matter, a smaller amount of helium (around 23 per cent) and tiny amounts of lithium and beryllium (just a fraction of a per cent). This general distribution of matter continues to this day, with the universe containing much more hydrogen and helium than anything else (if we ignore mysterious substances called dark matter and dark energy; see Spotlight at the end of the chapter).

But even when atomic matter first appeared, it wasn't distributed equally over the whole universe. Instead some areas of the universe contained more matter than others. This was a result of random fluctuations that occurred during the inflationary period. These fluctuations introduced variations in the distribution of the newly created matter that persisted as the universe expanded, eventually leading hydrogen and helium to form slightly denser clumps in certain regions.

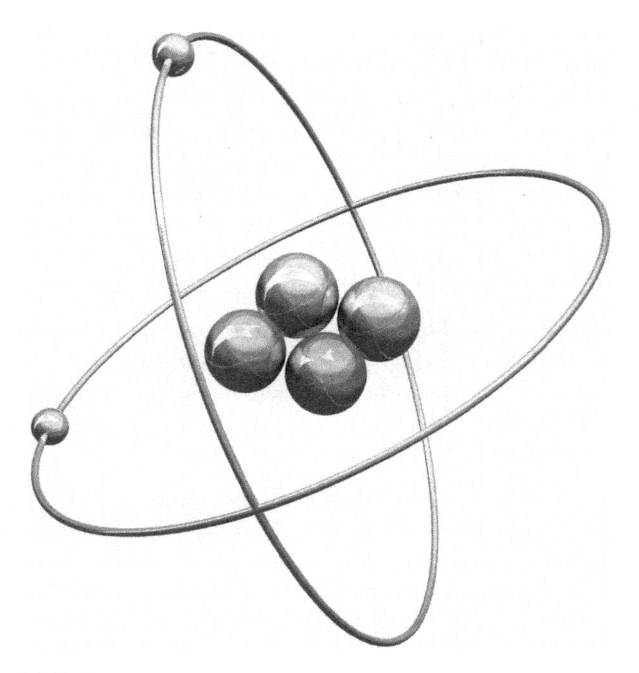

Figure 1.1 Helium atom

The force of gravity

This clumping tendency was then exacerbated by the force of gravity, which is the third of the four fundamental forces of nature (the fourth is the weak force, which is involved in some forms of radioactive decay). Gravity acts between bodies with mass, causing bodies with greater mass to attract bodies with lesser mass. As the clumps of hydrogen and helium possessed more mass than the surrounding regions, they attracted more matter to them, increasing their mass and thus attracting even more matter.

Eventually, after hundreds of millions of years, these clumps of hydrogen and helium formed the first stars and galaxies, separated by vast expanses of empty space. Finally, we had a universe that was similar to the one we inhabit today.

But still, this is really just the beginning of the story.

Spotlight: Over to the dark side

We may like to think of atomic matter as ordinary, making up us and everything we see around us. But atomic matter might not actually be very ordinary at all, perhaps accounting for less than 5 per cent of the 'stuff' of the universe. The vast majority of the universe appears to be made up of two mysterious substances that we know very little about: dark matter, which is thought to make up 27 per cent of the universe, and dark energy, thought to make up just over 68 per cent.

At the moment, we only have indirect evidence that these substances exist at all. Dark matter has been inferred from the movement of galaxies, because this movement seems to require more gravity than can be provided by the ordinary matter in these galaxies. This additional gravity is provided by dark matter.

Dark energy reared its head when astronomers were trying to pinpoint exactly how fast the universe is currently expanding, by monitoring distant exploding stars. Unexpectedly, this revealed that the rate of expansion is actually increasing, with astronomers postulating that this increase is being driven by a dark energy that permeates the entire fabric of the universe.

Initially, dark matter was thought to be exactly that: ordinary matter that was too dark to see in the depths of space, such as lone planets and the cores of dead stars. But scientists now think such objects only account for a small proportion of dark matter, with the majority attributed to one or more particles that we haven't encountered yet, dubbed weakly interacting massive particles (WIMPs). We haven't encountered these WIMPs because, as their name suggests, they probably don't interact much with ordinary matter except via the force of gravity, making them difficult to detect.

If anything, dark energy is even more mysterious. All this mystery raises the distinct possibility that dark matter and dark energy don't exist at all, and that we may need to find other explanations for our curious astronomical observations.

Key ideas

- The universe burst into being just under 14 billion years ago, in a process known as the Big Bang.
- The observable universe currently extends for around 46 billion light years in all directions and is expanding at 73 km a second.
- Atomic matter may account for less than 5 per cent of the 'stuff' of the universe, with two mysterious substances called dark matter and dark energy making up the vast majority.
- There are four fundamental forces of nature – gravity, and the electromagnetic, strong and weak forces.
- All atoms comprise a nucleus of protons and neutrons surrounded by a cloud of orbiting electrons.

Dig deeper

Butterworth, Jon, *A Map of the Invisible: Journeys into particle physics* (London: Windmill Books, 2018).

Singh, Simon, *The Big Bang: The most important scientific discovery of all time and why you need to know about it* (London: Harper Perennial, 2005).

We are all stardust

According to the British psychedelic rock band Hawkwind, space is deep. It's also cold, dark and mainly empty. But even in the depths of space, lone atoms occasionally meet and react with each other, joining together to form molecules.

Cosmic rays

For despite its cold and dark reputation, space is actually suffused with ultraviolet (UV) light and streams of subatomic particles and atomic nuclei known as cosmic rays, which both come from stars. These knock electrons off hydrogen, helium and oxygen atoms, forming positively charged ions that are primed to take part in chemical reactions (see Spotlight), even at temperatures as low as –263°C.

Spotlight: Ions

In Chapter 1, we learned that atoms don't possess a charge, because they contain equal numbers of negatively charged electrons and positively charged protons.

Sometimes, however, atoms can lose or gain extra electrons. If they lose one or more electrons, then they possess more protons than electrons and so become positively charged ions, known as cations.

If they gain one or more electrons, then they possess more electrons than protons and so become negatively charged ions, known as anions.

Oppositely charged ions can come together to form molecules. For example, table salt (NaCl) is created when sodium cations (Na^+) combine with chlorine anions (Cl^-).

These reactions take place when the ions crash into each other or into atoms of carbon, iron, nitrogen and silicon, reacting to form simple molecules such as carbon monoxide, water and silicates (various combinations of oxygen and silicon, particularly SiO_4).

Silicates are essentially tiny grains of sand and provide a solid surface onto which the various other molecules can freeze, forming an icy coating. This brings a whole host of these simple molecules into close enough proximity to react together, with the energy again provided by UV light and cosmic rays. These reactions can build up more complex molecules such as methanol, ammonia and formaldehyde.

If these ice-coated grains are then warmed by passing close to a star, the heat provides sufficient energy for these slightly more complex molecules to react together. This generates even larger and more complex molecules. These include: simple sugars such as glycolaldehyde; acetic acid, which is the main constituent of vinegar; and amino acetonitrile, which is related to the amino acids that are a central component of all life on Earth (see Chapter 4).

The formation of new elements

But wait a minute, where did all this oxygen, carbon, nitrogen, iron and silicon come from? In Chapter 1, we learned that the Big Bang generated loads of hydrogen, quite a lot of helium and a tiny amount of lithium and beryllium. How did the universe end up with the other 88 naturally occurring elements, producing 92 elements in total? Each of these elements possesses characteristic numbers of protons and electrons, and by joining together to form molecules, they produce all the matter we see around us today.

Well, the simple answer is that most of these other elements were forged in the fiery furnaces of stars under conditions of extreme violence.

The first stars lit up around 300 million years after the Big Bang, as gravity caused the vast clouds of hydrogen and helium to clump together into galaxies and then to collapse into individual stars. As a cloud collapses, its core becomes more and more compressed and therefore hotter and hotter. As a result, atoms of hydrogen and helium (we can ignore the lithium and beryllium) start slamming into each other.

At around 50,000°C, the atoms slam into each other with such ferocity that the collisions strip away their electrons, forming a mixture of hydrogen and helium nuclei surrounded by a gas of electrons. But the collapse carries on regardless: the hydrogen and helium nuclei keep smacking into each other and the heat continues to rise.

At 10 million degrees, the single protons that make up hydrogen nuclei hit each other with enough force to fuse together. But this force also causes one of the protons to decay into a neutron and so the end result is the nucleus of a deuterium atom, which is an isotope of hydrogen (see Spotlight). Thus, the process by which new elements are formed, known as nucleosynthesis, has begun, although the heat is still too low for helium nuclei to get involved.

Spotlight: Isotopes

In the famous periodic table, every element has its place. But in the real world things aren't quite so clear cut, because many elements exist in a number of different forms, known as isotopes.

The isotopes of an element all have the same number of electrons and protons, but different numbers of neutrons. So, a standard hydrogen atom has a nucleus made up of just one proton, but there are also two other isotopes of hydrogen: deuterium, with one proton and one neutron; and tritium, with one proton and two neutrons.

Different elements have different numbers of isotopes, from one to 10 (for tin). Although an element's isotopes tend to behave in similar ways, the rate at which they take part in chemical reactions can differ, with some important consequences.

Carbon has two main isotopes, known as carbon-12 and carbon-13, which have nuclei consisting of six protons and either six or seven neutrons (hence the numbers). Life on Earth is based on carbon, but life prefers to use the lighter carbon-12 over carbon-13. So, finding an abundance of carbon-12 in rocks can indicate that life existed when the rocks formed, whether on the early Earth or perhaps even on alien planets.

Now deuterium nuclei are flying around along with the hydrogen and helium nuclei. Deuterium nuclei start colliding with protons and fusing together, forming an unusual type of helium nucleus comprising two protons and one neutron. When two of these nuclei collide, they fuse together to become a standard helium nucleus comprising two protons and two neutrons, ejecting the two spare protons.

The helium nucleus is the end point of this suite of nucleosynthesis reactions, which release a huge amount of energy in the form of gamma rays. During their passage through the star, these gamma rays turn into other forms of electromagnetic radiation (see Chapter 17), including visible light, and generate an outward force that counters the collapse. Nevertheless, this force is not yet sufficient to prevent the core from collapsing further.

NUCLEOSYNTHESIS AND THE FORMATION OF STARS

At 25 million degrees, however, these nucleosynthesis reactions release enough electromagnetic radiation to halt the collapse and a stable, light-emitting star is born. During the remainder of its lifetime, the star will burn the store of hydrogen in its core as fuel, transforming it into helium and releasing electromagnetic radiation in the process. Outside the core, though, hydrogen doesn't burn, because it never gets hot enough.

The extent of a star's lifetime depends on its size. The largest stars (around 60 times larger than our Sun) are the shortest lived, burning through the hydrogen in their cores in just 60 million years and shining over 100,000 times brighter than our Sun. In contrast, the smallest stars (just one-tenth the size of our

Sun) should live for over 800 billion years and shine only one thousandth as brightly. Our middle-aged Sun should shine for around 10 billion years.

At the end of a star's life, having used up all the hydrogen fuel, its core begins to collapse again. This causes the temperature to rise even higher, triggering successive rounds of nucleosynthesis reactions. At 100 million degrees, the helium nuclei in the core start fusing together to produce carbon nuclei. (For the pedants out there, the three elements between helium and carbon in the periodic table – lithium, beryllium and boron – are not synthesized within stars. They are instead produced by cosmic rays careering into elements such as carbon and nitrogen in the depths of space, in a process known as cosmic ray spallation.)

As the core heats up, so do the outer layers of the star, triggering nucleosynthesis reactions in those layers that were hitherto too cold. All these nucleosynthesis reactions now taking place in the core and outer regions generate an enormous amount of electromagnetic radiation. This temporarily halts the collapse of the core and inflates the outer layers of the star, causing the star to expand by a factor of 100 to form a red giant.

But the core soon starts to contract again, as the star uses up each of its nucleosynthesis fuels, turning up the heat and triggering a new round of reactions. Inexorably, the temperature in the core rises from 100 million degrees to 6 billion degrees, synthesizing in turn the 21 elements in the periodic table from oxygen to iron. At this point, the whole nucleosynthesis process grinds to a halt, because iron is too stable to take part in any further reactions.

Without any nucleosynthesis reactions to prop it up, the core collapses until it can collapse no more, squeezing the iron nuclei and electrons into one huge nucleus. The core then explodes in a shower of neutrons and subatomic particles that rip the star apart, transforming it into a roaring fireball called a supernova that for a few months shines brighter than a whole galaxy.

As they tear through the star, the neutrons and subatomic particles collide with all the elements from carbon to iron now being produced via nucleosynthesis in the outer layers. This builds up the remaining 66 naturally occurring elements, which are then all flung far out into the depths of space.

Only the largest stars – those at least 20 times larger than the Sun – are able to go the full hog, synthesizing the complete set of

87 new elements. Stars ten times larger than the Sun only get as far as synthesizing silicon in their cores, but do still explode as supernovae.

Stars three times larger than the Sun only get as far as carbon, while the Sun will only ever get as far as helium. And rather than their cores exploding violently, they gently expire by exhaling huge flows of protons and electrons known as stellar wind. That's not quite the end, though, because all stars experience some form of afterlife, which is occasionally even more dramatic than their life (see Spotlight).

Spotlight: From white dwarfs to black holes

Once a star has exploded as a supernova or exhaled its last stellar wind, it leaves behind a charred remnant of highly compressed matter. Exactly what form this remnant takes depends on the size of the original star. Our Sun will leave behind what is known as a white dwarf, in which just over 50 per cent of the Sun's current mass will be squeezed into a similar volume to the Earth.

Stars between 10 to 25 times larger than the Sun leave behind a remnant consisting entirely of neutrons, known as a neutron star, just a few miles in diameter and containing up to three times the mass of the Sun. But it is stars more than 25 times larger than the Sun that meet the ultimate fate; so massive that the collapse of their core cannot be halted, they give birth to the most feared of all celestial bodies – a black hole.

Also known as a singularity, this is an infinitely dense region of space with such immense gravity than not even light can escape its pull. The point-of-no-return surrounding a black hole is known as an event horizon, close to which time slows and beyond which all known laws of physics break down.

The size of the event horizon depends on the mass of the black hole, which will increase if it is pulling in, or 'feeding' on, nearby stars or gas clouds. The largest black holes are thought to exist at the centre of many galaxies, including our Milky Way. Known as supermassive black holes, these giants can be millions to billions of times the mass of our Sun, and in 2019 scientists managed to capture the first image of one of them.

Since the first stars appeared 12 billion years ago, countless generations have been born and died. Billions of stars exploding in a huge continuous firework display that blasts the 87 elements throughout the universe, with the largest stars ejecting more matter than is found in 20 Suns. Despite this, hydrogen and helium still make up the vast majority (98.1 per cent) of the detectable matter in the universe. Carbon, nitrogen and oxygen are the next most abundant, making up 1.4 per cent, with the remaining 87 elements bringing up the rear on 0.5 per cent.

Spreading through space, these elements mingle with the clouds of hydrogen and helium left over from the Big Bang to form the dense molecular clouds that are the cradles of new generations of stars.

Spotlight: Sir Fred Hoyle, 1915–2001

Many of the details of how elements form in stars were worked out in the 1940s and 1950s by a British astrophysicist named Fred Hoyle, based on his knowledge of nuclear physics.

Later in his career, Hoyle went slightly off the scientific rails by arguing that simple life didn't originate on Earth but was carried here by comets. Known as panspermia, this idea is a step too far for most scientists and not supported by firm evidence. He also claimed that Earth was still being bombarded by deliveries of extra-terrestrial bacteria and viruses, triggering cancers and global pandemics.

His other claims to fame were coining the term 'the Big Bang' – although he always thought this theory of how the universe began was incorrect – and writing several science-fiction novels.

These dense molecular clouds, which can stretch for hundreds or thousands of light years, are also where the majority of the interesting interstellar chemistry takes place. This is because, as their name suggests, these clouds contain a greater concentration of material than is found in more diffuse regions of space.

But 'dense' is a relative term in the depths of space, because the concentration of material in a molecular cloud corresponds to the best vacuum that can be produced by scientists on Earth. Nevertheless, molecules and atoms collide or freeze onto silicate grains much more regularly in such clouds than in more diffuse regions of space.

The clouds also afford a degree of protection against the harsh environment of space, where UV light and cosmic rays can destroy molecules just as easily as they can create them. Outside of dense molecular clouds, only robust molecules such as carbon monoxide and a group of large, chicken-wire-shaped carbon-based molecules known as polycyclic aromatic hydrocarbons (PAHs) are able to survive for extended periods. On entering molecular clouds, the UV light and cosmic rays quickly lose much of their energy, allowing them to promote chemical reactions without destroying the results.

Using radio and infrared telescopes, scientists are continually discovering new molecules in these molecular clouds (see Chapter 17). At the time of writing, around 200 different molecules have been detected, including all those mentioned at the start of this chapter.

Some of the most complex molecules, such as ethyl formate, are detected in dense molecular clouds that have just given birth to a new star, such as the cloud known as Large Molecule Heimat. This supports the idea that such complex molecules are produced via the heating of ice-coated grains.

But laboratory experiments that involve heating simulated ice grains covered with molecules such as methanol, water, ammonia and carbon monoxide have generated even more complex organic molecules. These include a molecule called hexamethylenetetramine, which can spontaneously form amino acids when exposed to acid.

No amino acids have yet been conclusively detected in dense molecular clouds, perhaps because molecules as complex as amino acids are difficult to identify. But they have been found in a type of meteorite known as a carbonaceous chondrite, along with many other biologically important molecules. Carbonaceous chondrites are thought never to have been exposed to intense heat and so probably reflect the composition of the dense molecular cloud that gave birth to our solar system.

Similar biologically important molecules have also been detected in comets. These collections of dust and ice formed at the far edge of our solar system and are also thought to retain the composition of the parent molecular cloud.

So, a propensity for life may have been imprinted on our solar system right from the start.

Key ideas

- Space is suffused with ultraviolet light and cosmic rays that can induce interstellar atoms and molecules to react together.
- Most of the 88 elements that weren't produced by the Big Bang are forged in the fiery furnaces of stars, in a process known as nucleosynthesis.
- A star will burn the store of hydrogen in its core as fuel, transforming the hydrogen into helium and releasing electromagnetic radiation in the process.
- The extent of a star's lifetime depends on its size: our middle-aged Sun should shine for around 10 billion years.
- After using up all its fuel, a star expires to leave behind one of three celestial bodies: a white dwarf, a neutron star or a black hole.

Dig deeper

Chown, Marcus, *The Magic Furnace: The search for the origin of atoms* (London: Vintage, 2000).

Tyson, Neil Degrasse, *Astrophysics for people in a hurry* (New York: W. W. Norton & Co., 2017).

3

Recipe for a solar system

To make a tasty solar system, first take a fresh dense molecular cloud. This should mainly consist of hydrogen and helium left over from the Big Bang. But it should also contain a pinch of all the other naturally occurring elements spewed out by nearby supernovae, as well as simple molecules such as carbon monoxide, water and silicates formed by chemical reactions between these elements.

Then expose this dense molecular cloud to the shock waves produced by a nearby supernova, which will trigger the collapse of regions of the cloud under the force of gravity. Then wait a few hundred thousand years, at which point the temperature at the centres of these collapsing regions will have risen sufficiently to trigger nucleosynthesis, heralding the birth of new stars.

The collapse will also cause the new stars to rotate faster and faster. And being fairly massive, they will start to drag the rest of the collapsing regions around with them. Where once there was a molecular cloud, now there will be a whole host of new stars, each orbited by huge, flat expanses of rotating dust and gas extending for billions of kilometres. These are known as circumstellar discs.

Our solar system

Then just leave to simmer for around 100 million years, during which time these circumstellar discs will naturally turn into mouth-watering solar systems.

At least, that is the cosmic recipe that scientists think led to the formation of our own solar system around 4.5 billion years ago. Our solar system now comprises the four inner rocky planets (Mercury, Venus, the Earth and Mars) and the four outer gas giants (Jupiter, Saturn, Uranus, Neptune) separated by the asteroid belt (see Figure 3.1). Far out beyond the orbit of Neptune, there's also a tiny icy planet called Pluto, but whether it should actually count as a planet is a matter of some controversy.

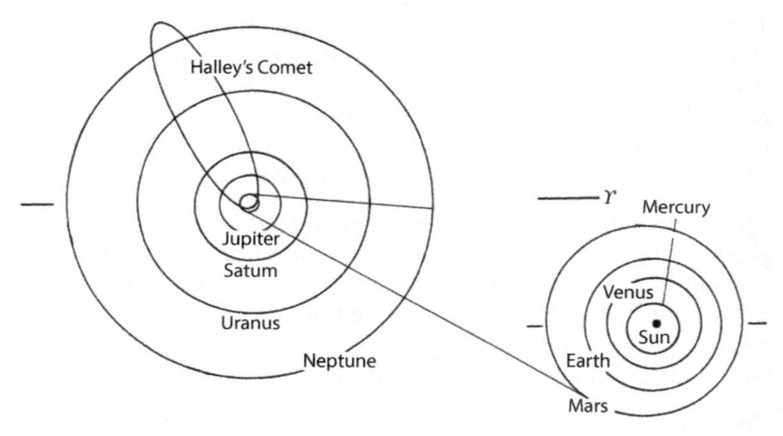

Figure 3.1 The solar system

Beyond Pluto is a large orbiting collection of icy bodies known as the Kuiper belt. Consisting mainly of comets, the Kuiper belt also contains small icy planets of a similar size to Pluto. Hence the controversy over whether Pluto should actually count as a proper planet at all. Or whether, along with the other small icy planets, it should be classified as a dwarf planet, which has been its official designation since 2006.

There is also an even larger collection of comets called the Oort cloud. Whereas the Kuiper belt probably contains millions of comets orbiting in a flat ring, the Oort cloud is thought to contain 100 billion comets orbiting as a quasi-spherical heap.

When considering distances in the solar system, kilometres are really too small a unit of measurement, but light years are too big.

So, astronomers tend to measure distances in terms of astronomical units (AU), which equals the distance from the Earth to the Sun (around 150 million kilometres). Using this measure, Mercury orbits at a distance of 0.4AU from the Sun, Jupiter orbits at 5.2AU and Neptune orbits at 30.1AU. The Kuiper belt extends from about 30AU to 55AU, while the Oort cloud extends from 2,000AU to 50,000AU or more.

THE TRANSFORMATION OF A CIRCUMSTELLAR DISC INTO OUR SOLAR SYSTEM

Astronomers also think they have a pretty good understanding of the process that transformed the circumstellar disc orbiting the young Sun into our solar system. The gas in this disc would have consisted mainly of hydrogen and helium, with some oxygen, carbon and nitrogen thrown in, while the dust would have consisted mainly of silicates of iron and magnesium with some additional grains of carbon, PAHs and metallic iron.

In total, the dust and gas would have equalled at most 10 per cent of the mass of the Sun, with gas accounting for 98.5 per cent and the dust accounting for the remaining 1.5 per cent. Initially, turbulence within the disc kept the gas and dust mixed together. But after the young Sun had finished forming, this turbulence would have subsided and the dust would have begun to settle out, in the same way that sand in a glass of water will quickly settle out if you stop stirring it. This took place over a period of just a few thousand years, after which the circumstellar disc consisted of a flat disc of dust surrounded by a thicker disc of gas.

Due to a combination of the heat generated as the dust and gas were pulled into the circumstellar disc and the heat generated by the young Sun itself, the temperature of the disc now began to increase, with regions of the disc close to the Sun getting hotter than those further away. And it is this range of temperatures along the disc, stretching from hot at the centre to cold at the edges, that dictated what our solar system would eventually look like.

ICY COATINGS

In Chapter 2, we learned that in the depths of space simple molecules such as water and carbon dioxide readily freeze onto tiny silicate grains, forming an icy coating. This brings the simple molecules in close enough proximity to react together, forming more complex molecules such as methanol, ammonia and formaldehyde.

The dust grains present in the circumstellar disc around the young Sun will have possessed such icy coatings, but the precise fate of those coatings depended on where a grain was located in the disc. Closer in to the Sun than the current orbit of Jupiter (5.2AU), the temperature of the circumstellar disc would have been above freezing, which in space is around –50°C. Thus, the icy coatings on the grains in the inner portion of the circumstellar disc would soon have evaporated away, leaving behind bare silicate grains. Beyond about 5.2AU, known as the snow line, the temperature of the circumstellar disc never got above freezing and so the silicate grains kept their icy coatings.

THE BIRTH OF JUPITER AND OTHER PROTOPLANETS

As the circumstellar disc orbited around the young Sun, the silicate grains began to stick together, forming small conglomerations that gradually became bigger as they swept up more and more grains, like a snowball rolling down a hill. After around a million years, this process had formed around 30 protoplanets varying from Moon-sized to Mars-sized within the snow line.

Beyond the snow line, however, the protoplanets built up much faster, simply because ice-coated silicate grains are slightly more massive than bare silicate grains. On top of this, the water and other molecules that evaporated from the silicate grains within the snow line condensed back onto grains as soon they reached the snow line, adding even more mass to the icy coatings.

As a result, after a million years a protoplanet with a mass greater than ten Earths had built up at the snow line. This protoplanet was now massive enough to start pulling all the gas and other matter in the surrounding area towards it under the influence of gravity. Within another million years, this process had bulked up the protoplanet to 30 times the mass of the Earth and surrounded it with an enormous gaseous layer equivalent to almost 288 times the mass of the Earth. Jupiter was born.

A similar process built up the three other gas giants, but it was increasingly less effective. This was because Jupiter had already captured a large proportion of the gas in the circumstellar disc, while the rest of the gas was now being blown away by a strong solar wind. So, there was less gas for Saturn, Uranus and Neptune to wrap around themselves. Whereas Jupiter is 318 times more massive than the Earth, Saturn is just 94 times more massive, and Uranus and Neptune are less than 20 times more massive.

In the warmer environs within the snow line, the rocky protoplanets never grew anywhere near large enough to acquire a huge gaseous atmosphere. Their existence was also more violent, as they were subject to repeated collisions, which either built them up or destroyed them. Eventually, after about one hundred million years, this demolition derby resulted in the formation of four rocky planets, as well as a collection of debris known as the asteroid belt.

The formation of the Earth

The largest of these rocky planets was the Earth, which like the other rocky planets mainly comprised a mixture of silicates and iron. As well as building up the Earth, the continual collisions turned the Earth's surface into a huge sea of molten lava. The intense heat at the surface caused the interior of the planet to heat up, melting both the silicates and the iron. Because iron is denser than silicates, the molten iron began to flow down through the silicates to the centre of the Earth, eventually forming a molten iron core that exists to this day (see Chapter 11).

Towards the end of this process, the Earth suffered a huge collision, in which a protoplanet slightly larger than Mars caught it a glancing blow. This collision destroyed the protoplanet and blasted a huge chunk of the Earth into space, where it formed an orbiting ring. Over a period of 10 million years, this rocky debris came together to form the Moon.

After that huge collision things calmed down a bit for the Earth, until countless comets and asteroids started slamming into it, in a heavy bombardment that continued for the next 500 million years. Consisting of agglomerations of ice-coated grains, these comets had formed out beyond the snow line and initially travelled in circular orbits around the Sun.

But the growth of the gas giants disturbed their orbits, flinging the comets all over the place. Some were flung completely out of the solar system, and some ended up forming the Oort cloud. Others were sent speeding towards the inner solar system, where they ran the risk of colliding with the four rocky planets.

For the Earth, however, the heavy bombardment had an up-side, because these comets and asteroids delivered much of the water that filled up the oceans and much of the carbon-based material that eventually spawned life.

These cooking instructions were pieced together by astronomers from observations of our solar system, which until recently was all they had to do go on. That all changed a few decades ago, when the first planet orbiting another star was detected. Now, astronomers have thousands of solar systems to study and this is revealing that some of their cooking instructions may have been slightly different (see Spotlight).

Spotlight: Super-Earths and hot Jupiters

In 1992, astronomers detected the first planet outside our solar system, or exoplanet, confirming that our solar system was not unique. What initially started out as a trickle of further discoveries over the next 15 years or so turned into a flood in 2009, with the launch of the Kepler space telescope, which was specifically designed to hunt for exoplanets. Kepler ran out of fuel in 2018, but over its nine-year lifetime it detected around 2600 exoplanets. Together with detections made by ground-based telescopes, this means that, at the time of writing, there are now around 4000 confirmed exoplanets.

They have proved so common that astronomers think almost every star in the Milky Way – and maybe even the universe – probably has at least one orbiting planet. But even though this means our solar system is certainly not unique, it may still be unusual. Because the two most common types of exoplanet detected to date are quite unlike anything found in it.

One type is known as super-Earths, rocky planets that range in size from the Earth to Neptune (which has a diameter four times that of the Earth), and the other type is known as hot Jupiters, huge gas giants that orbit very close to their star. Earth may be the largest rocky planet in our solar system, but most super-Earths are much larger, while current theories of solar system formation don't allow gas giants to form so close to a star. This implies either that those theories are, at best, incomplete or that gas giants often migrate towards their star after formation. Either way, this all raises a lot of questions.

Some of these will hopefully be answered by the next generation of planet-hunting telescopes, both Earth- and space-based. These include the James Webb space telescope, due to be launched in 2020, which as well as finding exoplanets will also be able to detect light reflected from them. This will allow it to probe the chemical composition of exoplanet atmospheres and perhaps even produce images of them.

Spotlight: Hunting high and low

Astronomers have come up with two different ways to detect exoplanets. First to be developed was the radial-velocity method, used to detect the initial clutch of exoplanets, but it has subsequently been usurped by the transit technique, which is responsible for the Kepler space telescope's impressive haul.

The radial velocity method detects exoplanets by the slight wobble they induce in a star's movement as they orbit around it, causing the star to oscillate towards and then away from the Earth. This movement is reflected in the wavelength of light from the star that reaches Earth (see Chapter 17). When the star is moving away from the Earth, its light is stretched into longer, redder wavelengths (known as a redshift) and when moving towards the Earth its light is squeezed into shorter, bluer wavelengths (known as a blueshift). These shifts thus indicate the presence of one or more planets.

The transit technique detects exoplanets by the slight dip in the star's brightness that occurs when they pass in front of it. Although this is more efficient than the radial velocity method, allowing multiple stars to be monitored simultaneously, it does require the star and the Earth to be in the same plane, so that the exoplanet passes between them.

These two techniques are complementary rather than competitors, as they can each reveal different information about an exoplanet. The radial velocity method can reveal the mass of an exoplanet, as more massive planets produce larger wobbles, while the transit method can reveal the size of an exoplanet, as larger planets block more starlight. From the mass and size of a planet, astronomers can calculate its density and thus determine whether it's made of rock or gas.

Despite their differences, both techniques are most effective at detecting large planets that orbit close to a star, as they cause the largest wobbles and greatest dips. This might help explain why astronomers have so far detected so many super-Earths and hot Jupiters.

Key ideas

- Our solar system formed 4.5 billon years ago, out of a huge, flat expanse of dust and gas known as a circumstellar disc orbiting the young Sun.
- Our solar system now comprises the four inner rocky planets (Mercury, Venus, the Earth and Mars) and the four outer gas giants (Jupiter, Saturn, Uranus and Neptune) separated by the asteroid belt.
- The snow line, beyond which the temperature never reached high enough to melt the icy coating on silicate grains, marks the transition between the rocky planets and the gas giants.
- The Moon was created when the young Earth collided with a protoplanet slightly larger than Mars.
- Astronomers have so far discovered around 4,000 exoplanets, of which the most common types are super-Earths and hot Jupiters.

Dig deeper

Cohen, Andrew and Cox, Brian, *The Planets* (London: William Collins, 2019).

Schrijver, Karel, *One of Ten Billion Earths: How we learn about our planet's past and future from distant exoplanets* (Oxford: Oxford University Press, 2018).

Life begins

Around 3.8 billion years ago, the Earth was not a particularly pleasant place.

The period of heavy bombardment had only just finished, leaving the ground and sea boiling at temperatures of over 100°C. Volcanoes were everywhere, constantly spewing out huge volumes of steam and carbon dioxide, as well as sulphur dioxide, hydrogen and nitrogen. This produced a thick, suffocating atmosphere consisting mainly of carbon dioxide, with smaller amounts of methane, carbon monoxide, hydrogen and nitrogen. Incessant acid rain lashed through these heavy skies, scarring the rocks on the ground. And all the while, comets and meteorites were still occasionally slamming into the Earth.

But still, in the middle of this Dante-esque vision, life may well have made its first appearance. Or at least rocks of this age in Greenland, which are some of the oldest known, contain evidence that life existed when they formed. Now this evidence, consisting of a slight overabundance of carbon-12 over carbon-13 (see Chapter 2), is hotly disputed. This is because the rocks have been extensively modified by heat over the ages, making any interpretation of their chemical composition very tricky.

But if life wasn't around just at that point, it probably didn't wait much longer to show up. Rocks from Australia dating back almost 3.5 billion years contain microscopic structures that look very much like the fossilized remains of tiny bacteria, while some fossilized colonies of such bacteria, known as stromatolites, have also been dated to around the same time.

So just a few hundred million years after the end of the ferocious bombardment of meteorites and comets, which would have snuffed out any life that tried to get going, the seas of the Earth were probably full of microbes. Now this is a surprisingly short amount of time to go from lifeless to teeming and suggests that given the right conditions simple life arises without too much trouble, perhaps within as little as 20 million years. But exactly how it arose is still very much open to question.

LUCA

The problem is that there's no way to see that far back into the mists of time. Early bacteria may have left fossilized remains, but the very first life form, termed LUCA (last universal common ancestor), didn't. So, in piecing together what actually happened, scientists have been following two lines of inquiry.

The first approach involves discovering those 'right conditions' that led to life. By working out the kind of chemical compounds that would have existed on the early Earth and the kind of reactions that could have taken place between them, scientists are trying to work out how these chemicals came together to form life. The second approach involves looking at the basic components of current life and then working backwards to determine how they may have developed.

The first can be seen as a kind of bottom-up approach, going from lifeless to life, and the second as a kind of top-down approach, going from life to lifeless. The hope is that these two approaches will eventually meet in the middle to produce a plausible route from lifeless chemicals to seas teeming with microbes. The semblance of such a route is now just beginning to appear.

A definition of life

But before we start tramping along this route, perhaps it would be useful to define what we actually mean by life. While it may seem fairly simple to distinguish life from non-life – dogs are alive, stones aren't – producing a formal definition is actually quite difficult.

Perhaps the most obvious criterion for life is being able to reproduce. If something is alive, then it must be able to produce copies of itself, which may or may not be identical. But this criterion is not sufficient on its own, because crystals are able to grow and produce identical copies of themselves if placed in salt solutions. And no-one would argue that crystals are alive.

To the ability to reproduce we need to add the ability to evolve. For something to be alive, the copies it produces of itself need to be able to change gradually across the generations in response to environmental factors. Evolution will be explained more fully in Chapter 5, but it is responsible for transforming the simple microbes that floated in the seas of the early Earth into the multitudinous variety of life forms that exist today (and have ever existed).

THE ESSENTIAL PROPERTIES OF LIFE

Now we have a basic definition of life, we can start to pick out its essential properties, which should have been possessed by the very earliest forms. The most essential of these essential properties is that all life on Earth is constructed from molecules containing carbon and hydrogen, known as organic molecules. And if there's one aspect of the early Earth that scientists are fairly certain about, it's that there was an abundant supply of organic molecules.

This supply came from a number of different sources. For a start, the meteorites and comets that pummelled the early Earth brought with them huge amounts of different organic molecules. Organic molecules were also naturally produced in the thick atmosphere, as a result of reactions powered by lightning and sunlight. Finally, organic molecules would have been produced under the sea at hydrothermal vents. These are cracks in the Earth's crust that emit large amounts of hot water from underground. This water has been heated by circulating around molten rocks, and so contains carbon dioxide and various other chemical compounds from the rocks, which can react together to form organic molecules.

So, any body of water on the early Earth, from oceans to puddles, would have been awash with organic molecules. Heated by the Sun or volcanoes or the continuing flow of hot water from hydrothermal vents, these molecules would have started reacting together. Now in most cases this would simply have resulted in an almighty mess, but occasionally a stable system of reactions would get going, in which the product of one set of reactions would feed into the next set. In this way, order would have emerged from chaos for the first time.

How were amino acids joined together?

One family of organic molecules that would have been available on the early Earth is amino acids, which consist of various combinations of carbon, hydrogen, oxygen and nitrogen surrounding a central carbon atom. Scientists know this because they have detected a wide range of amino acids in both meteorites and comets. Furthermore, experiments simulating the kind of reactions that would likely have taken place in the early Earth's atmosphere and around hydrothermal vents also produce amino acids (see Spotlight).

Spotlight: Early Earth in a flask

Replicating one of the likely stages in the formation of life is actually fairly easy, at least for an academic chemist. Simply fill a flask with a mixture of methane, ammonia, hydrogen and water vapour (to simulate the early atmosphere), and repeatedly apply heat and electricity (to simulate volcanoes and lightning).

When a young US chemist called Stanley Miller did this in 1953, he found that after a few days the water turned brown. Analysing this water, he detected a whole range of complex organic molecules, including several amino acids found in proteins. An analysis of stored samples from this experiment with more advanced instruments in 2008 revealed an even greater range of amino acids.

Scientists now think that the early Earth's atmosphere probably contained much more carbon dioxide than methane or ammonia, and when the same experiment is performed with this mixture of gases far fewer organic molecules are produced. Nevertheless, Miller's experiment goes down in history as the first to show that biologically important molecules can spontaneously be produced from a simple mixture of gases.

This is interesting, because amino acids are the building blocks of proteins, and proteins are the very cornerstone of all current life on Earth. Life is built from proteins and works due to proteins. And all the thousands and thousands of proteins utilized by all the life forms on Earth are made up of long chains of just 20 different amino acids.

The challenge then is to find a mechanism that naturally joins amino acids together into these long chains. Today, this is done within cells in a complex process that involves lots of different components, based on instructions coded in the cell's genes. How could proteins have been produced without all this cellular apparatus? Furthermore, amino acids can't simply join together randomly; they have to join together in a set order to produce a set protein with set properties. How could that have happened without any kind of biological guidance?

RNA AND ROCKS

Well, one theory is that it didn't initially happen with proteins at all, but rather with a molecule known as RNA (ribonucleic acid), which is very similar to the DNA that makes up genes. Like DNA, RNA is made up of organic molecules known as nucleotides (see Chapter 6), which were also probably lying around on the early Earth (or at least their component parts were).

The advantage RNA has over proteins is that RNA theoretically contains the instructions for replicating itself (via a process that will be explained in Chapter 6), offering a way for the same molecule to be made over and over again. What is more, scientists have discovered that certain RNA molecules can perform some of the same functions as proteins, including speeding up chemical reactions.

So perhaps RNA came first, forming the basis for the earliest forms of life and creating what is known as the 'RNA World', before eventually synthesizing the proteins that subsequently took over. But that still leaves a big gap between the first stable reactions and the dawning of the RNA World; what could possibly have bridged that gap?

The answer may well be as simple as 'rocks'. The idea is that the ultimate products of the stable reactions, perhaps including several nucleotides, would have regularly washed into the pores of rocks on the shore of some sea or lake, or around hydrothermal vents at the bottom of oceans. Here, the molecules would have joined together to produce more complex molecules, with the surface of the rocks acting as a template, aligning the molecules and helping them to link together in set ways. The pores might even have acted like cell walls, concentrating molecules inside them. As the surface of rocks don't change very rapidly and the same organic molecules were repeatedly washed onto them, the rocks acted like an assembly line, continuously producing the same complex molecules.

With this process repeated all over the early Earth for tens or hundreds of millions of years, it's perhaps not too surprising that eventually a range of RNA molecules (or initially perhaps slightly simpler versions) were produced. Then all it would take was for some of these molecules to become incorporated within fatty bubbles in the water and, hey presto, LUCA appears.

Now, of course, this is all speculation. Although scientists have found a certain amount of laboratory evidence for their theories, including chemical routes for producing biologically important molecules such as ribose, an essential component of RNA, and RNA molecules that can replicate other RNA molecules, they are still a long way off finding a plausible route from non-life to life.

The 'RNA World' hypothesis is also not the only game in town. Rather than find non-biological ways to produce the molecules currently utilized by life, such as RNA, some scientists argue that the focus should be on finding networks of stable reactions. In this view, what's important is not the identity of the molecules, but that they form stable networks of reactions that can persist over time, perhaps also within the pores of rocks. This persistence would allow the networks to become steadily more complex, eventually leading to life, but perhaps to a life that initially utilized a whole different range of molecules.

Once life got going, however, there is abundant evidence for what happened next.

Spotlight: Life elsewhere

The study of the origin of life on Earth would receive a tremendous boost if life was ever discovered elsewhere in the solar system. Not only would this strongly suggest that life is almost inevitable given the right conditions, rather than just a lucky fluke, but it could also help reveal whether life always requires the same set of molecules or can function perfectly well with a different set.

Mars has always been a prime candidate for harbouring extra-terrestrial life, both in fiction and reality. Although cold and dry now, it was much warmer and wetter in the distant past, billions of years ago. At this point, liquid water, an essential condition for life, may well have flowed across its surface. Even today, liquid water could still exist under the Martian surface, where microbial life may thrive.

In 2018, a group of astronomers announced the discovery, based on radar readings from an orbiting probe, of a 20-km-wide lake of liquid water buried 1.5 km beneath Mars's surface, close to its southern polar ice cap. A similar subterranean lake in Antarctica, called Lake Vostok, which has been isolated from the rest of the world for tens of millions of years, is known to contain plenty of life.

But other, more surprising candidates are also beginning to appear. These include Jupiter's moon Europa, Saturn's moon Enceladus, and the dwarf planets Ceres, in the asteroid belt, and Pluto, out beyond Neptune. Several lines of evidence suggest that they may all possess huge, planet-encompassing subterranean oceans of liquid water, warmed by heat from their interiors.

Only in the case of Enceladus, however, has this liquid water actually been seen, as plumes jetting from its surface. These were witnessed by the Nasa spacecraft *Cassini* during its exploration of Saturn. In 2008, *Cassini* even flew through one of the plumes, collected some of the water and analysed it, finding that it contained organic molecules, such as methane, considered to be likely precursors of life.

Key ideas

- Life may have first appeared on Earth around 3.8 billion years ago.
- There would have been an abundance of organic molecules on the early Earth, from sources such as comets and meteorites, atmospheric reactions and hydrothermal vents.
- Early life may have been based on RNA rather than proteins, or on an entirely different set of molecules.
- Rocks may have acted as templates or vessels for stable reactions that produced more complex molecules.
- Several bodies in the solar system may contain large expanses of liquid water that could harbour life.

Dig deeper

Lane, Nick, *The Vital Question: Energy, evolution, and the origins of complex life* (London: W. W. Norton & Co., 2016).

Pross, Addy, *What is Life? How chemistry becomes biology* (Oxford: Oxford University Press, 2016).

5

Evolution and extinction

As soon as the first living, reproducing microbes began to appear in the seas of the early Earth, evolution by natural selection kicked into gear.

Now, it can be argued that a form of chemical evolution was operating even in the non-biological era, eventually leading to the development of life. But it was only once life finally appeared that evolution really had something to get its teeth into.

For as we saw in Chapter 4, one of the defining features of life is the ability to reproduce. But reproducing all the components of a living organism, even a microbe, is a lot more complicated than reproducing a simple non-biological structure such as a crystal, which merely consists of repeating a physical pattern. With this added complexity comes an increased chance of making a mistake and such mistakes (or mutations) drive evolution.

Beneficial mutations

Often the mutations will be small enough to have little or no noticeable effect on the microbial offspring. But occasionally the mutations will result in the microbial offspring being slightly different to the parent. Sometimes this difference will be detrimental to the offspring, resulting in its quick death, but occasionally it will be beneficial, providing the new microbe with an advantage over all the other microbes in the vicinity.

As a result, this new, upgraded microbe will be able to out-compete its fellows for resources, allowing it to grow faster and reproduce more. The same will happen for its own upgraded offspring, which will therefore come to dominate the microbial population, until eventually every microbe is upgraded. This process continues over and over again with each new generation, causing the microbes to change and evolve as they gradually accumulate beneficial mutations.

This is how evolution by natural selection, as first postulated by the British naturalist Charles Darwin over 150 years ago (see Spotlight), essentially works. Reproduction is the centre-piece of the whole process. It provides both the mechanism by which organisms change and the mechanism by which any beneficial changes propagate through the population, because organisms with such changes reproduce more than those without.

Spotlight: Charles Darwin, 1809–82

For Charles Darwin, 2009 was a momentous year. It marked 200 years since his birth and 150 years since the publication of his groundbreaking book *On the Origin of Species*, in which he first set out his theory of evolution by natural selection.

The theory came out of Darwin's years of meticulous study of a whole range of different organisms, from dogs to worms. Perhaps most important was his trip to South America on board the HMS *Beagle* from 1831 to 1836. As part of this trip, he stopped at the Galapagos Islands and noticed how each individual island contained its own unique species of animals and plants.

Darwin first came up with his theory in 1838, but mindful of the likely impact he held off publicizing his ideas for almost 20 years. Only when a young naturalist called Alfred Russell Wallace wrote to him expressing similar thoughts did he go public.

As he predicted, the effects were seismic. He had come up with the first plausible mechanism for explaining the entire diversity of life on Earth that didn't require the hand of a creator.

The role of the environment

The other main player is the environment, in the form of food sources, competitors, predators and the physical habitat, because this dictates which changes are beneficial and which aren't. Beneficial changes are those that enhance an individual's ability to operate in its specific environment – allowing it to gain food, compete against its fellows, avoid predators and safely navigate its habitat – thereby giving it a better chance of reproducing.

This means that changes that are beneficial in one specific environment could well be detrimental in another. For example, the ability to withstand hot temperatures would be very useful for a microbe living near a boiling hydrothermal vent, but far less useful for one living in a frozen lake.

So as soon as the first living, reproducing microbes appeared in the seas of the early Earth, evolution began changing them. The first major evolutionary leap occurred pretty quickly, when some of the microbes evolved the ability to photosynthesize. In other words, powered by the energy in sunlight, they were able to take the abundant carbon dioxide in the atmosphere and react it with the water in which they floated to produce organic molecules, specifically simple forms of sugar.

CYANOBACTERIA AND PHOTOSYNTHESIS

The first organisms known to be able to do this are called cyanobacteria (which are still around today) and they formed the stromatolites that provide some of the earliest indications of life (see Chapter 4). This means that photosynthesizing cyanobacteria may have first appeared around 3.5 billion years ago.

The ability to photosynthesize represents an incredibly important advance. By generating their own nutrients, microbes no longer had to rely on an external source, which lifted a lot of restrictions. The newly photosynthesizing microbes were able to grow more rapidly and travel much more widely, allowing them to colonize the whole planet.

THE RISE IN OXYGEN LEVELS

There was also a less immediate but perhaps even more important consequence. Like all photosynthesizing organisms, cyanobacteria produce sugars by combining the carbon in carbon dioxide with the hydrogen in water, but this leaves a lot of unwanted oxygen, which photosynthesizing organisms simply release into the atmosphere.

Now, a single cyanobacterium releases a miniscule amount of oxygen in its lifetime, but countless cyanobacteria continuously producing oxygen over millions and millions of years gradually transformed the Earth. The released oxygen reacted with rocks, dissolved in the oceans and then built up in the atmosphere. Whereas 3.5 billion years ago oxygen accounted for just 0.1 per cent of the Earth's atmosphere, by 2 billion years ago it had risen to 3 per cent (it now accounts for around 20 per cent).

This rise in oxygen levels then kicked off the next major evolutionary leap. Oxygen may be essential for most current life on Earth, but it is an extremely reactive element that damages biological structures given half the chance. So, the early microbes were forced to evolve mechanisms to deal with the higher concentrations of oxygen.

Some microbes went one step further and actually made use of the oxygen. They utilized its high level of reactivity to break down organic molecules more fully than they were previously able to, releasing more energy and inventing respiration.

A few of the microbes started to seek refuge within hardier colleagues. These microbes eventually merged to become one, creating larger and more complex microbes collectively known as eukaryotes. This process also resulted in the first major split in Earth's life forms. Those eukaryotes that housed photosynthesizing microbes became photophytes, which are the ancestors of all plants, while those that housed respiring microbes became protozoans, which are the ancestors of all animals.

This all occurred around 2 billion years ago. Life then had to wait another 500 million years for the next evolutionary leap, but it was well worth it. For around 1.5 billion years ago, the eukaryotes invented sex.

Sexual reproduction

Now, sex may have a number of benefits (see Chapter 8), but the one that concerns us here is that it super-charges evolution. By mixing the genetic traits from two individuals, sexual reproduction produces offspring that always differ from their parents. Whereas asexual organisms can only evolve at the whim of random mistakes, sexual organisms have change built into their very being.

Sex greatly expanded the range of opportunities available to evolution. In fact, it was probably essential for the next major evolutionary leap, which took life over a particularly high hurdle.

Life may have first appeared on Earth 3.5 billion years ago, but 2.5 billion years later it still consisted of numerous single-celled microbes floating in the sea.

THE FIRST MULTI-CELLED ORGANISMS

Only at this point did the single-celled eukaryotes start to come together to form larger groups and colonies. At first, the eukaryotes making up these groups would have been all the same. Eventually, however, they started to specialize and differentiate into different varieties, resulting in the formation of the first proper multi-celled organisms around 750 million years ago.

At first, these were mainly types of jellyfish and sponges, but over the following 200 million years life gradually diversified, producing some highly unusual creatures (see Spotlight). Towards the end of that period, organisms with shells and exterior skeletons, such as trilobites, first started to appear (see Figure 5.1). Just over 100 million years later, the seas were full of a huge array of different organisms, including the first fish with backbones. At around the same time, plants began to migrate from the seas and onto the land, leading to a vegetation explosion.

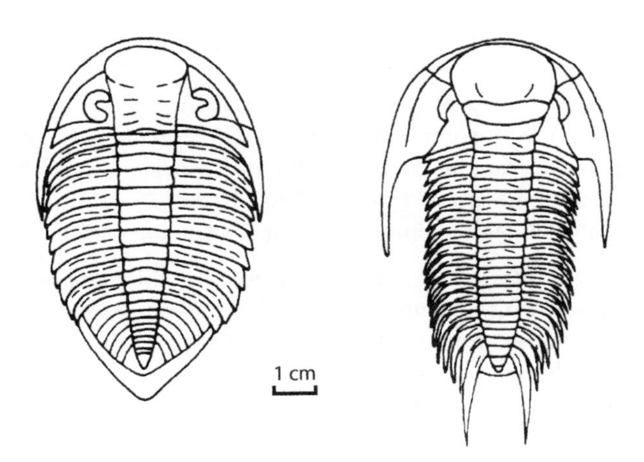

Figure 5.1 Trilobites

With the land now covered in edible plants, animals also began to leave the sea. Scorpions were first, around 390 million years ago, followed by a whole variety of other invertebrates that soon evolved into insects. Then, 30 million years later, the first fish crawled on their fins up onto the land, evolving first into amphibians and then reptiles.

By 225 million years ago, some of the reptiles had become quite large, ushering in the age of the dinosaurs, which lasted for the next 160 million years. During this time, the world also witnessed the appearance of the first mammals, around 200 million years ago, and the first flowering plants, around 75 million years ago.

MASS EXTINCTIONS

Evolution has both winners and losers. So alongside the appearance of all these new types of life, others were going extinct. Now extinction is a process that runs alongside evolution: individual species go extinct all the time as a result of their habitat changing or being out-competed by more effective species. On top of this, however, large numbers of different species occasionally become extinct in a relatively short period of time, such as a few hundred thousand years.

Such events are known as mass extinctions. They can vary considerably in severity and extent, but five mass extinctions have been severe enough to affect almost all life on Earth. The exact causes of these five mass extinctions have still not really been pinned down, but probably involved one or more of the following: impact by a meteorite or comet, major volcanic activity lasting for millions of years and ice ages.

As well as the immediate damage, major volcanic activity and meteorite or comet strikes also cause more long-lasting difficulties. Meteorites can throw huge volumes of debris into the atmosphere, blocking out the Sun and killing off vegetation, while volcanoes can release massive amounts of carbon dioxide, which cause the Earth to heat up through global warming (see Chapter 23). Ice ages cause damage through falling temperatures and shrinking oceans.

The third and fifth of these mass extinctions are the best known. The third one, which occurred around 250 million years ago, is the largest ever mass extinction, killing up to 90 per cent of all living organisms. The last one, which occurred 65 million years ago, famously killed off the dinosaurs.

The emergence of humans

But mass extinctions are not all bad news, because they create the conditions for a whole new round of evolution. Habitats that were until recently home to numerous species are now available for new species to colonize. Indeed, only 5–10 million years after a mass extinction takes place, the diversity of life often exceeds that before the mass extinction.

To give an example close to home, the end of the dinosaurs proved to be a tremendous opportunity for the mammals, which they grasped with all paws, going on to spread over the entire Earth. And as they spread out, they evolved, until eventually, around 200,000 years ago, modern humans first strode onto the scene.

Spotlight: Any old fossils

Scientists have pieced together the history of life on Earth from the fossil record. Fossils are essentially mineral-based imprints of organisms that lived millions of years ago. They are produced when newly dead plants and animals are covered in sediment, either silt or mud in water, or sand and volcanic ash on the land.

As more and more material is laid on top, the sediment surrounding the dead organism is compressed until it hardens into sedimentary rock like sandstone and limestone. Organic material such as flesh quickly decomposes away, leaving behind the mineral components of bones and teeth or gaps that fill with mineral deposits. In this way, the original organism is replaced by a mineral facsimile in the rock.

Although you're unlikely to find a whole dinosaur skeleton, smaller fossils can be found wherever there are sedimentary rocks. For instance, some of the first fossils were found along the Dorset coast around Lyme Regis, which remains a fossil hot spot.

Spotlight: Like nothing else on Earth

Scientists may have pieced together the history of life on Earth from the fossil record, but some fossils have proved difficult to fit into that history, because they look totally unlike anything else. Most of these bizarre fossils come from the earliest stages of that history, specifically from geological periods known as the Ediacaran and the Cambrian.

The Ediacaran period extends from 635 million years ago to 542 million years ago, and plays host to the earliest fossils of multi-celled animals. This period has been referred to as 'the Garden of Ediacara' because the animals that existed then were soft-bodied and mostly stationary, probably feeding on microbes rather than each other; a simple, peaceful time. The Cambrian period follows straight after the Ediacaran period and continues until 485 million years ago, and it is here that claws and teeth for attack and shells and spikes for defence, as well as limbs and eyes for both, first appear.

Ediacaran animals include *Parvancorina*, which looks like an anchor covered in biological tissue, and *Dickinsonia*, which looks like a segmented bath mat. There is also *Kimberella*, which looks like a hovercraft and may well have crawled slowly over the bottom of ancient seas.

With more attributes to play with, some of the Cambrian animals look even stranger. *Opabinia* has five eyes, a flat, lobed body, a fan-shaped tail and a long proboscis with evil-looking claws at the end, like the jaws that shoot out of the mouth of the alien in the titular films. *Hallucigenia* is essentially a worm with legs, as well as a host of spikes on its back and several tendrils extending from its head end.

Scientists are still unsure whether many of these creatures are bizarre ancestors of existing forms of animal – the latest thinking is that *Hallucigenia* is an early form of arthropod (the grouping that contains insects and spiders) – or evolutionary experiments that reached a dead-end.

Key ideas

- Evolution by natural selection, in which beneficial mutations allow organisms to produce more offspring, is the process that transformed seas of simple microbes into the vast array of life on Earth today.
- Five mass extinctions have been severe enough to affect almost all life on Earth, including one around 65 million years ago that killed off the dinosaurs.
- In the first 2.5 billion years of its existence, simple microbial life evolved photosynthesis, respiration and sex, and split into the ancestors of all animals and plants.
- The first multi-celled organisms appeared around 750 million years ago, and life first began to migrate from the seas onto land around 420 million years ago.
- Modern humans first appeared around 200,000 years ago.

Dig deeper

Brannen, Peter, *The Ends of the World: Volcanic apocalypses, lethal oceans and our quest to understand earth's past mass extinctions* (London: Oneworld Publications, 2018).

Fortey, Richard, *Life: An unauthorised biography* (London: Flamingo, 1998).

Part Two

Way of all flesh

6 Life in sequence

Over the years, scientists have occasionally fallen into the trap of thinking that they're on the cusp of a complete understanding of some area of science just as that understanding is comprehensively blown out of the water.

It happened to physics in the early years of the twentieth century, when scientists first discovered that the subatomic world could not be explained by classical physics, ushering in the new field of quantum mechanics (see Chapter 25). Then, in the early years of the twenty-first century, it happened to biology.

The proteins that make up life

The problem lies in the relationship between DNA and proteins. As we saw in Chapter 4, proteins are the cornerstone of all life on Earth. Life is built from proteins and works due to proteins.

Each of the tens of thousands of different proteins utilized by life is made up of just 20 different types of amino acid, which are joined together to form long molecular chains. The longest of these chains contains over 20,000 amino acid molecules, although the average protein only contains a few hundred, but the important point is that the sequence of amino acids in these chains is unique for each protein.

The amino acids are able to link together into chains because their ends are essentially sticky. This stickiness takes the form of a chemical bond, which attaches each amino acid molecule to its immediate neighbours, producing long, flat chains.

But that's not all, because some of the amino acids can also link up with colleagues in other, more distant parts of the chain via other chemical bonds. These chemical bonds pull the flat chains of amino acids into a complex three-dimensional structure, like a form of molecular origami, with the precise shape of the structure determined purely by the sequence of amino acids. And it is this precise shape that determines what a protein is able to do.

STRUCTURAL AND FUNCTIONAL PROTEINS

There are basically two types of protein: structural and functional. As their names suggest, structural proteins build up biological material such as muscle, cartilage and hair; functional proteins, on the other hand, ensure that life stays functioning. Most importantly, they do this by greatly speeding up (otherwise known as catalysing) the numerous chemical reactions that are essential for life, such as breaking down food. Such catalytic proteins are known as enzymes and it is their precise shape that allows them to function in this way.

Proteins are produced by, and mainly act within, cells. This is true both for single-celled microbes such as bacteria and for each of the hundred trillion cells that make up our bodies (see Chapter 7). And the instructions for producing all the thousands of different proteins are contained within each cell's genes. If proteins are life's workers, then genes are life's managers.

The DNA molecule

Genes are able to direct the production of all these proteins because they are made up of DNA (deoxyribonucleic acid). A single molecule of DNA consists of three components: a simple sugar known as ribose that has lost one of its oxygen atoms (hence the 'deoxy' prefix); a small phosphorus-containing molecule known as a phosphate group; and a very special molecule known as a nucleotide base.

The ribose and phosphate group act as the 'backbone' of DNA, essentially providing support for the nucleotide base and linking the numerous DNA molecules together into long strands. But the really important part of a DNA molecule is the nucleotide base, which can be one of four different molecules: adenine (A), guanine (G), cytosine (C) or thymine (T).

So, in a similar vein to proteins, a strand of DNA consists of the four different nucleotide bases joined together in a specific sequence. This similarity is no coincidence, because the sequence of nucleotide bases is directly related to the sequence of amino acids in a protein, acting as a kind of coded instruction for producing proteins.

And that's not the only similarity with proteins. For in the same way that chemical bonds can form between different amino acids, they can also form between the nucleotide bases. Importantly, though, adenine can only form a bond with thymine and guanine can only form a bond with cytosine.

The upshot of all this is that DNA doesn't exist in the cell as a single strand but as a double, conjoined strand, in which the sequence of nucleotide bases is always matched with a kind of 'mirror-image' sequence. So, A always matches with T and G always matches with C, all along the conjoined strand (see Figure 6.1).

Not only is this set-up very chemically stable, but it also provides a simple way to produce a copy of any DNA sequence, as is required for cell division (it is mistakes in this copying process that produce the mutations that drive evolution; see Chapter 5). What's more, it gives DNA its characteristic double-helix shape, with the two strands wrapping around each other, as first elucidated by the British molecular biologist Francis Crick and the US biochemist James Watson in 1953.

As such, double-stranded DNA is shaped much like a twisted ladder, with the alternating ribose and phosphate groups forming each of the two vertical struts and the sequence of matching DNA bases forming the horizontal rungs.

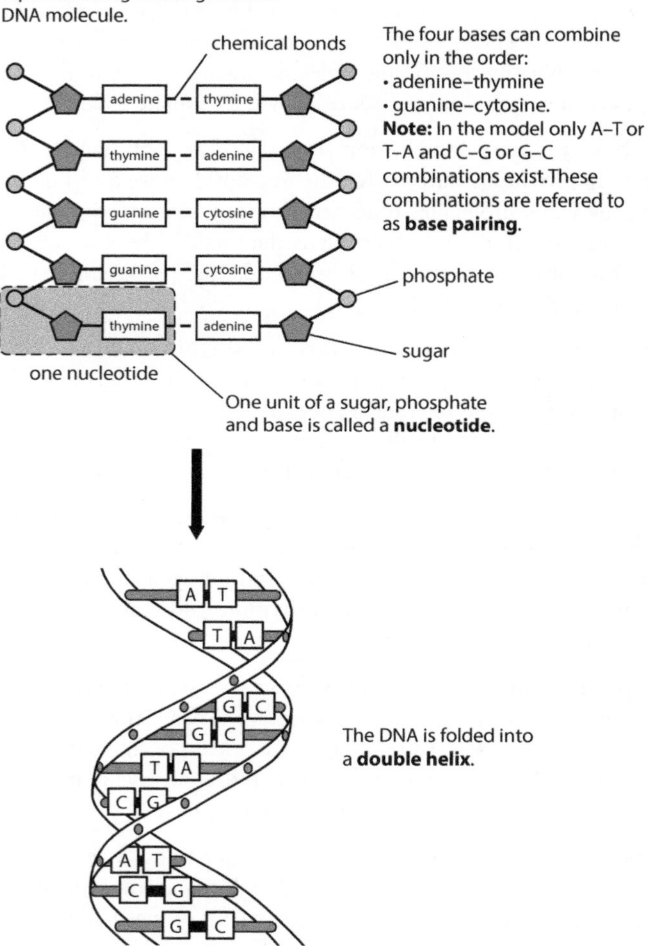

Figure 6.1 DNA

DECIPHERING THE DNA CODE

So, the question now is: how exactly does the sequence of nucleotide bases relate to the sequence of amino acids in a protein? Or, in other words, exactly how does this code work?

Well, first off, the long sequence of DNA bases is grouped into smaller collections known as genes. If DNA bases are letters, then genes are words: more specifically, they are effectively the names of different proteins.

To get at these genes, an enzyme splits the two DNA strands apart. Once a gene is exposed, another enzyme then creates a mirror image of that gene, but this mirror image is created from RNA (ribonucleic acid) rather than DNA.

RNA is very similar to DNA, except that its ribose section isn't missing an oxygen atom, it exists as a single strand and it contains the nucleotide base uracil (U) rather than thymine. In all other respects, however, the generated RNA strand acts as a direct mirror-image copy of the gene.

This RNA strand then travels away into the main body of the cell, where it is eventually captured by a cellular structure known as a ribosome. This clamps around the RNA strand and then proceeds to travel along it, reading the sequence of nucleotide bases.

Specifically, it reads these bases in groups of three, known as codons, because each group of three bases corresponds to a particular amino acid. So, the sequence CAA corresponds to valine and the sequence UCG corresponds to serine. As it travels along the RNA strand reading each codon, the ribosome essentially brings out the corresponding amino acid from a nearby store and adds it to the growing protein (see Figure 6.2).

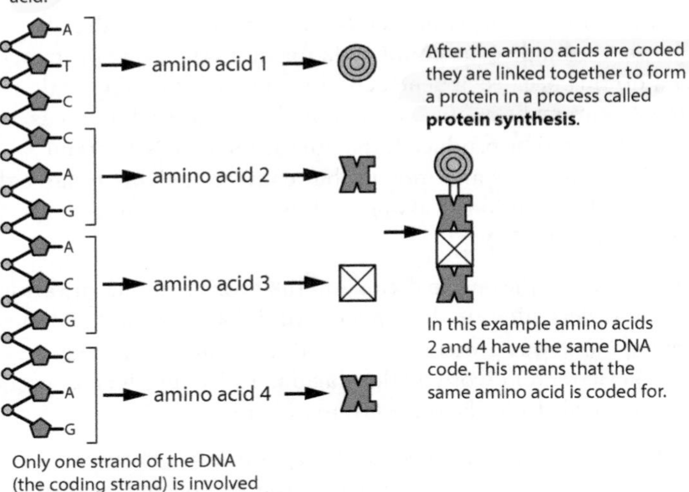

Figure 6.2 Making a protein

The Human Genome Project

So, there you have it, in all its elegant glory: genes contain the instructions for producing proteins, while RNA physically carries these instructions to the cell's protein-constructing machinery. Although proteins may do all the work, it is genes that are calling the shots. It is genes that dictate who we are and what we look like; genes that encapsulate life's master plan. Understand genes and you'll understand life.

That was the thinking behind the Human Genome Project, which aimed to determine the sequence of all the 3 billion pairs of nucleotide bases making up human DNA, known as the human genome. A draft sequence of the human genome was released in 2000, with a final version published in 2003. But rather than revealing humanity's remaining genetic secrets, this completed sequence simply highlighted how much scientists still didn't understand.

The problem was that, based on the protein production process detailed in Figure 6.2, scientists had assumed that there was a one-to-one relationship between genes and proteins, with individual genes coding for individual proteins. As such, many scientists estimated that the human genome would contain around 100,000 genes. In actual fact, it contains just over 20,500.

Exons and introns

This was a major surprise, even though scientists did already know that a single gene could code for more than one protein. It can do this because a gene actually consists of coding regions of DNA, known as exons, and non-coding regions, known as introns. When first produced, the mirror-image RNA strand contains both exons and introns, but enzymes quickly remove the introns and then join the exons back together to form a single coding piece of RNA.

The trick is that the enzymes can join the exons together in various different orders, altering the sequence of the RNA strand and thus allowing numerous proteins to be produced from the same gene. Using our letter and word analogy again, it's like producing several different words from the same group of letters.

Scientists knew this happened but thought it was a fairly rare occurrence. It now seems that such alternative splicing, as it is known, may occur in more than half of all human genes.

Switching genes on and off

On top of this, it appears that the genetic machinery is much more flexible than hitherto realized. Rather than being a static repository of information, a cell's DNA is continuously being chemically and structurally modified. This results in whole swathes of genes being switched on and off all the time, providing extremely fine control over protein production. Scientists knew this happened – it's why a single genome can produce lots of different types of cells (see Chapter 7) – but didn't appreciate that it was quite so dynamic. It means a cell's genetic machinery can quickly respond to changes in external conditions, such as reductions in nutrient availability.

Scientists are also discovering that unassuming RNA may be the real power behind the genetic throne. For one question that has always troubled scientists is why so much of the human genome appeared to be useless. Only about 3 per cent of our genome codes for proteins; the other 97 per cent, termed junk DNA, seemed to serve no purpose.

Scientists assumed that this junk DNA had just built up over the millennia, but it now turns out that most of this junk DNA is not junk at all. Instead, it codes for RNA strands that, rather than producing proteins, directly control when individual genes are turned on and off, termed non-coding RNA. This was confirmed in 2012, when a decade-long project called Encyclopaedia of DNA Elements (ENCODE) revealed that around 80 per cent of the human genome serves a clear purpose, and for the majority of the genome (over 70 per cent) that purpose is producing non-coding RNA.

As far as genes are concerned, scientists are discovering that it's not what you've got that counts, but how you use it.

Spotlight: Sins of the fathers

Not only have scientists recently discovered that the genetic machinery is much more dynamic and flexible than previously supposed, allowing it to respond quickly to changes in external conditions, but they have also found that these changes can be passed down the generations. What this means is that the conditions your parents and grandparents experienced when they were growing up can directly affect your health, growth and behaviour.

The idea that than an individual can pass traits acquired as a result of their experiences on to their offspring has always been anathema to scientists. In the early nineteenth century, a French naturalist called Jean-Baptiste

Lamarck proposed it as a mechanism for evolution; under this theory, giraffes evolved long necks because successive generations kept on stretching up to reach high leaves. But it was comprehensively scotched by Charles Darwin's theory of natural selection, by which giraffes with longer necks produce more offspring than those with shorter necks, and the discovery of genes. Now, however, this idea, known as epigenetics, is experiencing something of a renaissance.

While genes remain the sole means for passing on heritable traits, it turns out that the precise expression of those traits, as influenced by an individual's specific experiences, can also be passed on. One way that life experiences can alter genetic expression is via chemical modification of the genome, whereby specific molecules bind to genes to switch them on or off. It now appears that this pattern of chemical modification can be passed on to eggs and sperm, and thus to an individual's offspring and even to their offspring's offspring.

Over the past few years, scientists have uncovered several examples of this. One is that whether an individual experienced famine in their youth can influence the lifespan of both their children and grandchildren. Another is that rats exposed to stressful conditions produce more anxious pups, which remain anxious even if they aren't exposed to stressful conditions. While yet another is that when pregnant rats are exposed to nicotine, not only do their offspring develop abnormal lungs but so do their offspring's offspring, even if they have never been exposed to any nicotine.

Spotlight: Genes in sequence

Humans are just one of a large number of plants and animals that have had their genomes sequenced. Others include chimpanzees, dogs and maize. Although these organisms are all very different, their genes are made up of the same four nucleotide bases and can therefore be sequenced in exactly the same way.

Developed by the British chemist Fred Sanger in the 1970s, gene sequencing is an ingenious and elegant process. It involves creating numerous copies of a DNA strand, but stopping this copying process at a random point each time. Furthermore, the final nucleotide base in each copy is automatically tagged with a fluorescent compound that shines one of four different colours depending on the base.

Do this enough times and you'll be left with DNA fragments of every possible length, with each successive fragment one base longer than the previous one. Then all you need do is separate these fragments by their length, which can be done by essentially sieving them through a gel, and record the colour given off by the fluorescent tag at the end of each fragment. In this way, you can build up the sequence of the original strand (see Figure 6.3).

Advances in technology over the years have led to this technique, known as Sanger sequencing, becoming much faster, but they have also led to the development of rival techniques that are even faster, able to sequence a genome in less than an hour. One example is ion torrent sequencing, which sequences a genome by detecting the hydrogen ions released when a specific base is added to a growing DNA strand. Other, even faster sequencing techniques are in the pipeline, such as reading bases by passing a DNA strand through a tiny pore in a membrane.

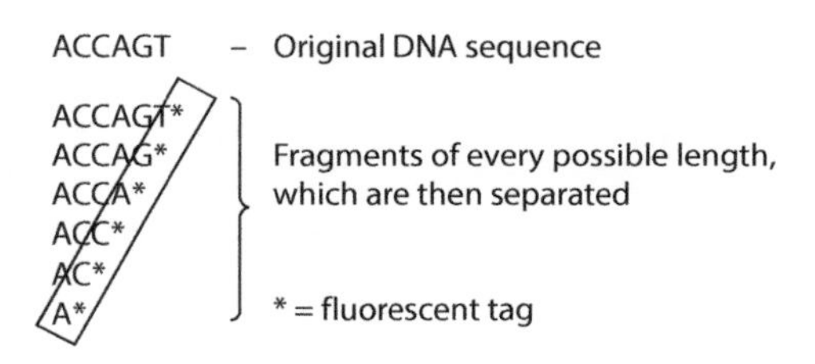

Figure 6.3 Gene sequencing

Key ideas

- Proteins are made up of 20 different types of amino acid, which are joined together to form long molecular chains.
- A DNA molecule contains one of four different nucleotide bases: adenine (A), guanine (G), cytosine (C) or thymine (T).
- In a cell, DNA exists as a double, conjoined strand, in which the sequence of nucleotide bases is always matched with a kind of 'mirror-image' sequence.
- The sequence of nucleotide bases in a gene relates to the sequence of amino acids in a protein.
- Only about 3 per cent of the human genome codes for proteins, while the vast majority (over 70 per cent) is responsible for producing non-coding RNA.

Dig deeper

Carey, Nessa, *Junk DNA: A journey through the dark matter of the genome* (London: Icon Books Ltd, 2015).

Rutherford, Adam, *A Brief History of Everyone Who Ever Lived: The stories in our genes* (London: Weidenfeld & Nicolson, 2017).

7

From the bottom up

OK, you might now be saying, I get that genes code for proteins, and that the rest of the genetic machinery is responsible for switching those genes on and off. But how does that process produce us, in all our unique glory? How does it result in all our different bodily tissues and organs; how does it ensure that we are different from the other 7 billion people on the planet; how does it ensure that humans are different from the millions of other species?

The genome

In answering this question, the first point to make is that just as humans have their own genome, comprising all their protein-coding genes and associated genetic machinery, so does every other species of plant and animal on the planet. And those genomes are all different, which is why humans, mice and oak trees all look and behave very differently. Furthermore, within each species that reproduces sexually, which includes most plants and animals, almost every individual member of that species has its own unique genome, which is why those individual members often look and behave differently.

Organisms that reproduce asexually, such as bacteria, produce offspring that are direct clones, possessing exactly the same mix of genes as their lone parent (barring any genetic mutations). Species that reproduce sexually can sometimes produce offspring with identical genomes (in humans, identical twins possess exactly the same set of genes). But most organisms with two parents are genetically unique.

GENOME VARIATION

Now, genomes differ more between different species than between members of the same species. So, on average, the genomes of different humans differ by only around 0.5 per cent (the Human Genome Project sequenced an amalgamated genome derived from several different people). A large number of these differences take the form of variations in a single nucleotide base within a gene – one person may have guanine while another will have thymine. These variations are known as single nucleotide polymorphisms (SNPs).

The differences between the genomes of different species are greater, with the degree of difference generally reflecting the level of similarity between the species. Human genomes differ from chimpanzee genomes by just 1.2 per cent, but differ from mouse genomes by around 15 per cent.

Still, a 15 per cent difference is not that much considering how different men are from mice. What it indicates is that the vast majority of the genes in an individual organism's genome are concerned with keeping that individual alive, rather than differentiating it from other organisms. Perhaps unsurprisingly, mice appear to need many of the same proteins as humans. Such a small difference also supports the growing realization that what's important is not the specific genes we possess but how we use them.

The second point is that all the cells that make up an individual organism contain exactly the same set of genes. This is why we can talk about a person's genome, even though that genome is actually located in each of the hundred trillion cells that make up our bodies. Because the genome is the same in every cell, every person effectively has a single genome.

DIFFERING NUMBERS OF CHROMOSOMES

In the eukaryotic cells that form all multi-celled organisms, this genome, comprising a long sequence of paired DNA molecules, is found within a membrane-bound organelle known as a nucleus. Because the DNA sequence is very long (the human genome comprises 3 billion pairs of nucleotide bases, which if stretched out would extend for 2 metres), it is split up and wrapped tightly around proteins known as histones to form numerous chromosomes.

Different species have different numbers of chromosomes, reflecting the amount of DNA in the nuclei of their cells. In general, 'more complex' species have more chromosomes than 'simpler' species, but that is not always the case. A mouse has 20 pairs of chromosomes, while a human has 23 pairs, but a guinea pig has 32 pairs. More evidence that it's not what you've got, but how you use it.

PAIRS OF CHROMOSOMES

Note here the mention of pairs of chromosomes; each of the chromosomes in a eukaryotic cell possesses an almost identical twin. This is because the cells of organisms produced by sexual reproduction gain a chromosome from each parent.

Most of the genes on these chromosomes are identical, or should be, as they code for proteins that are essential for life and are therefore the same in both parents. Indeed, having pairs of chromosomes turns out to be a pretty good defence against genetic defects. In many cases, if one copy of a gene works fine, then it doesn't matter if the other one is defective; it's only when both genes are defective that problems can arise.

This is the case with sickle cell anaemia – a genetic disorder that causes red blood cells to become deformed, reducing their ability to carry oxygen. Sickle cell anaemia is caused by a defect in a single gene that codes for part of a protein complex known as haemoglobin, but the genes on both chromosomes need to be defective for the disorder to manifest itself fully. If only one gene is defective then the person usually remains healthy; indeed, there is even a benefit to having a single copy of the defective gene because it seems to confer resistance to malaria.

DOMINANT AND RECESSIVE ALLELES

Some of the genes for less essential traits can differ slightly between the pairs of chromosomes; such genes are known as alleles (with the various alleles often a result of SNPs). One allele tends to be dominant while the others are recessive, which means that the dominant allele will always be expressed over the recessive alleles (see Spotlight).

Spotlight: Gregor Mendel, 1822–84

The notion that an organism's traits are passed on to its offspring in the form of discrete units [later termed genes], rather than all mixed together, was first shown by an obscure German monk called Gregor Mendel. Although at first no one, including Mendel himself, realized exactly what he had discovered.

By growing garden peas with various different traits in the garden of his monastery in the 1850s and 1860s, Mendel discovered that certain traits would appear in these plants in simple ratios. For instance, if he crossed green peas with yellow peas, then all the offspring were yellow peas. But if he crossed two of the yellow pea offspring, then three-quarters of their offspring were yellow peas and one-quarter were green peas.

We now know that the reason for this is that the yellow pea allele is dominant and the green pea allele is recessive, only being expressed when a pea possesses two green pea alleles. But it took another 35 years for scientists to realize the full implications of Mendel's findings, by which time he was dead.

Most human characteristics (or traits) are determined by interactions between multiple genes, but a few traits have been linked to a specific gene that can exist as two or more alleles. One example is whether you have free or attached ear lobes, with the free earlobe gene dominant and the attached earlobe gene recessive.

How the cell functions

OK, you may now be saying, I understand how our different genomes distinguish humans from mice, and me from you, but how does a single, unique genome produce all the different types of cell and tissue that make up the body of a single multi-celled organism. How does the same genome produce the brain, heart, bones and skin? To answer that, we first need to explore the cell a bit further.

The 'brain' of the eukaryotic cell is the gene-containing nucleus, but – in the same way that in addition to brains we need hearts, lungs, skin and various other organs to stay alive – the cell needs a whole host of other organelles to stay functioning. These organelles are housed outside the nucleus in a granular substance known as the cytoplasm, which makes up most of the volume of the cell.

These organelles include the endoplasmic reticulum, which forms a series of channels through which proteins and other biological molecules are transported around the cell. There are two types of endoplasmic reticulum: rough and smooth. Rough endoplasmic reticulum is covered with loads of protein-producing ribosomes, and thus produces proteins and then transports them around the cell. Smooth endoplasmic reticulum, on the other hand, produces fatty molecules known as lipids and transports those around the cell.

There is also the Golgi apparatus, which is involved in modifying, storing and transporting proteins and other biological molecules, especially if they are destined to be released from the cell. This involves passing the molecules through the cell membrane, which is the physical boundary of the cell.

The cell membrane consists of a double layer of lipid molecules embedded with proteins that act as gateways in and out of the cell. Some cells, including bacteria and plant cells, also have an outer cell wall, providing enhanced strength and protection.

There are also various enzyme-containing sacks known as lysosomes that digest nutrients brought into the cell, producing compounds that can then be further broken down to produce energy (see below) or used as the building blocks for other biological molecules. The cell is held together by the cytoskeleton, which is a scaffolding-like network of fibres.

Finally, there are important organelles called mitochondria. These use oxygen to break down simple sugars such as glucose, producing energy for the cell and generating carbon dioxide as a waste product. All the cells in our body use mitochondria to produce energy, which explains why we breathe air and exhale carbon dioxide.

OXYGEN AND CARBON DIOXIDE

Many plant cells also contain important organelles called chloroplasts, which are responsible for conducting photosynthesis. They use energy from the Sun to convert carbon dioxide in the atmosphere into simple sugars like glucose, producing oxygen as a by-product. Hence, life on

Earth forms a huge, mutually reinforcing cycle, in which animal cells rely on the sugars and oxygen produced by plants, and plant cells rely on the carbon dioxide produced by animals.

Different types of cells in the human body

So that is the make-up of an average cell, but there is no such thing as an average cell. The human body consists of around 220 different types of cell, which vary widely in their size, shape and function.

Some, such as the cells in the salivary gland, produce lots of proteins and enzymes, which they secrete into their external environment, and so contain a rich system of rough endoplasmic reticulum. Some, such as the muscle cells that make up the heart, contain large numbers of mitochondria, because they need lots of energy to keep the heart pumping. Fat cells, meanwhile, mainly consist of huge sacks known as vesicles crammed full of fat molecules.

Some cells, such as red blood cells, have lost their nucleus. Others produce molecules that are not found in any other cell of the body, such as the light-responsive pigments produced by the cells that make up the retina of the eye.

But all these widely different cells contain exactly the same set of genes, meaning that theoretically a heart cell can turn into a retina cell. Their differences are explained by the fact that different genes are turned on, or expressed, in different cells, producing a unique mix of proteins that construct and operate a specific type of cell. Every cell contains the gene for producing light-responsive pigments, but it is only turned on in the cells of the retina.

There are even some cells that contain only a half set of chromosomes. It is at this point, however, that we need to pause, turn down the lights, put on some romantic music and light some scented candles, because we are about to enter the murky but exciting worlds of sex and reproduction.

Spotlight: When microscopy goes super

We know all about cells and their various organelles and what they do thanks to microscopy. By peering through microscopes for over 300 years, scientists have gradually built up a detailed picture of cellular life. And this picture has become increasingly detailed over the years as microscopy technologies have advanced, allowing scientists to perceive ever smaller cellular features.

But there is a fundamental limit beyond which scientists cannot peer, known as the diffraction barrier, caused by the fact that light waves tend to interfere with each other at small enough scales. For visible light, this means that cellular features smaller or closer together than 250 nanometres (nm; billionths of a metre) cannot be resolved, which includes fat-containing vesicles and the microtubules that make up the cytoskeleton. These features can be resolved by electron microscopy, which uses beams of electrons rather than light to produce magnified images. But electron microscopy requires the sample to be placed in a vacuum, meaning it can't study live cells or active cellular processes.

Over the past 20 years, however, scientists have developed a suite of techniques for overcoming the diffraction barrier. These techniques are collectively termed super-resolution microscopy, and their developers were recognized with a Nobel Prize for Chemistry in 2014. The techniques work in several different ways, but they are all forms of fluorescence microscopy, in which samples are labelled with fluorescent compounds that emit visible light when excited with light at different wavelengths, often infrared. Like covering a sample with lots of tiny lightbulbs.

In conventional fluorescence microscopy, all these fluorescent labels emit light at the same time, which means labels and features closer than 250 nm can't be resolved from each other. But by cleverly controlling this fluorescence, often by getting the labels to emit light one at a time and then producing a combined image of all of them, super-resolution microscopy can resolve cellular features that are less than 50 nm apart.

Key ideas

- Every species of plant and animal on the planet has its own genome, and almost every member of a sexually reproducing species has its own unique genome.
- All the cells that make up an individual organism contain exactly the same set of genes, but different genes are expressed in different cells.
- In eukaryotic cells, genes are housed within several pairs of chromosomes, with each one of those pairs coming from a different parent.
- Eukaryotic cells contain various organelles, including the nucleus, the endoplasmic reticulum, the Golgi apparatus and mitochondria, surrounded by a cell membrane.
- The human body consists of around 220 different types of cell, which vary widely in their size, shape and function.

Dig deeper

Arney, Kat, *Herding Hemingway's Cats: Understanding how our genes work* (London: Bloomsbury Sigma, 2017).

Wolpert, Lewis, *How We Live and Why We Die: The secret life of cells* (London: Faber & Faber, 2010).

8

Getting it on

Before we get embroiled in sex, let's first deal with reproduction, because the two are by no means synonymous. Every living organism is able to reproduce – as we saw in Chapter 4, it is one of the defining features of life – but not every living organism has sex. To determine the difference between the two, we need to remain for a while at the level of the cell.

Cell reproduction and mitosis

In many cases, when a single cell wants to reproduce – whether one of the cells that makes up our body (although not all the cells in our body can reproduce) or a single-celled organism – it simply splits into two, creating a direct copy of itself. For eukaryotic cells, this process is known as mitosis. Bacteria divide by a similar process known as binary fission.

When a cell is not actively reproducing via mitosis, it is usually preparing itself for mitosis. This mainly involves the cell creating a copy of all the DNA in its nucleus, a process made fairly straightforward by the 'mirror-image' structure of DNA (see Chapter 6). In a human cell, this means a copy of each of the 46 chromosomes.

CHROMATIDS

Once this process is complete, each chromosome is accompanied by an exact copy of itself. The original chromosome and its copy are now termed chromatids and are physically joined together at a point at their centres.

As the cell starts to divide, each of the 46 chromatid twins are pulled apart and dragged towards opposite ends of the cell by long tubular structures known as spindles that traverse the cell. The cell then forms a membrane between its two halves and divides. One cell has become two, with each cell containing a full set of 46 chromosomes.

Meiosis

For the first 2 billion years of its existence, this was the only form of reproduction available to life on Earth. But around 1.5 billion years ago, a few eukaryotic cells came up with an alternative method. Instead of splitting into two, they experimented with splitting into four, in a process we now term meiosis (see Figure 8.1).

This is possible because eukaryotic cells contain pairs of chromosomes, one from each parent; so in human cells, the 46 chromosomes are actually 23 pairs of chromosomes (see Chapter 7).

As in mitosis, a cell planning to undergo meiosis first duplicates all the DNA in its nucleus, such that each chromosome is transformed into two identical chromatids. In a human cell, there are 23 pairs of chromosomes, with each chromosome now consisting of two identical chromatids, producing 92 chromatids in total ($23 \times 2 \times 2$).

At this point, however, something else happens, because each pair of chromosomes (with each chromosome comprising two identical chromatids) starts to exchange genes between themselves, termed 'crossing over'. This process creates 92 totally new chromatids, each of which contains a unique mix of genes.

Then, the 23 pairs of chromosomes are separated, with one pair pulled to each end of the cell, which then divides. Each of the daughter cells now reverts to standard mitosis, meaning that the two chromatids comprising each chromosome are pulled apart and dragged to the opposite end of each cell, which then divides. In this way, one cell becomes four and, crucially, each of the four offspring contains only one copy of each of the 23 chromosomes.

Here's where sex comes into the equation, because you may now be asking yourself what is the point of meiosis? If a single cell can repeatedly split into two by mitosis, why bother with a convoluted process such as meiosis, especially as it results in just half-a-cell, genetically speaking.

Well, the whole point of meiosis is that the resultant half-a-cell (known as a haploid cell) can now fuse with another haploid cell to form a cell with the full complement of paired chromosomes (known as a diploid cell). The advantage of this approach over mitosis is that it produces a genetically unique diploid cell. Rather than being a direct copy of a single parent, this diploid cell is a mixture of two parents, receiving half its genes from each. Furthermore, because of 'crossing over', each haploid cell produced from a single parent cell by meiosis will possess a unique mix of genes, allowing the same two parents to produce offspring with a variety of different traits.

Fertilization

So, sex is merely the fusing of two different haploid cells. But how does that fusion, or fertilization, take place? Well that is the million-dollar question and one to which evolution has come up with a myriad of different answers.

When the first eukaryotic cells had sex in their ancient oceans, they probably fused together before undergoing meiosis, as still happens today when certain single-celled organisms indulge in sexual reproduction. The advent of multi-celled organisms in the oceans changed things, however, because now meiosis was limited to certain cells and so had to happen before fusion.

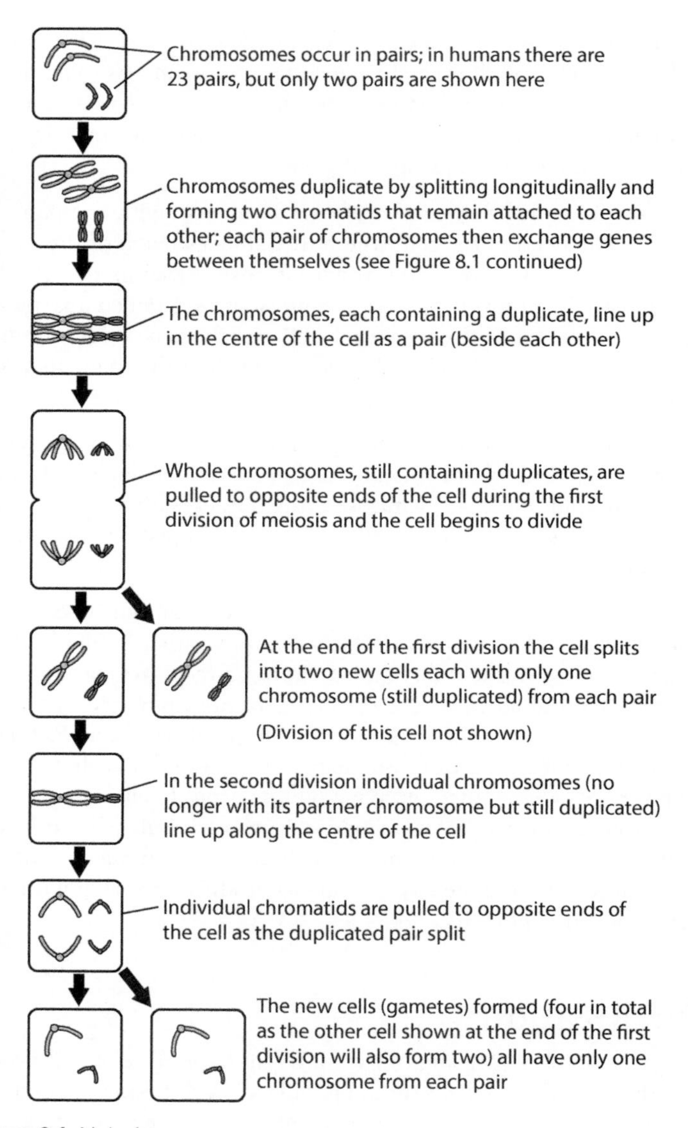

Figure 8.1 Meiosis

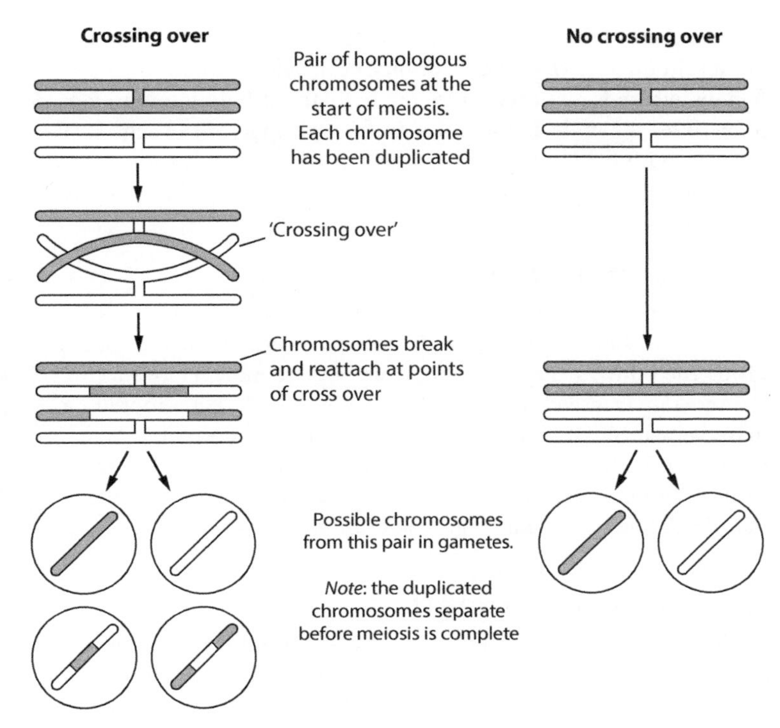

Figure 8.1 *(Continued)*

THE FUSION OF MALE AND FEMALE GAMETES

Furthermore, haploid cells differentiated into two types – one produced by males and one produced by females – and only male and female haploid cells, also known as gametes, could fuse. For animals, these gametes are sperm and eggs, while for plants they are pollen (male) and gametophytes or embryo sacs (female).

Initially, early marine creatures probably adopted a pretty basic sexual strategy, based around simply releasing their sperm and eggs into the water and hoping for the best. Indeed, some of today's marine creatures, such as sea urchins, still rely on this strategy.

The problem with this approach is its uncertainty; you never know whether your sperm or eggs have successfully merged with those of another member of your species. A slightly less chancy method is to make sure you only release your eggs and sperm at a propitious moment. For example, mussels and oysters, which feed by filtering nutrients from the water, only release sperm or eggs when they detect the converse gamete in the water.

Increased certainty of success can be gained if the male releases his sperm directly onto a collection of eggs already laid by a female, which is a strategy adopted by many fish and amphibians. Sometimes the male simply releases free sperm over the eggs, as is the case with fish, but other organisms such as crabs produce special packages of sperm known as spermatophores that the female rubs over the laid eggs.

INTERNAL FERTILIZATION

Alternatively, an organism can dispense with external fertilization entirely and shift to internal fertilization, where the fusing occurs inside the body of the organism, usually the female. Here success is almost guaranteed, which explains why it has been adopted by many marine creatures, including octopuses and dolphins. And because animal gametes can only merge when suspended in some form of fluid, it is the only form of fertilization open to organisms that don't live in or near large bodies of water.

Spotlight: Animal perverts

On the face of it, sex should come pretty far down the list of priorities for your average animal. Eating and avoiding being eaten should be much more important than getting your leg over, but that is not always the case. Some animals, such as male elephant seals, stop eating almost entirely during their mating season. Others, such as spiders and praying mantises, risk being eaten by their mate during, after and occasionally even just before sex (a male praying mantis can still have sex after losing its head to the female).

One of the ways evolution has pushed sex up the list of priorities is by making it hugely enjoyable. An offshoot of this is that many species like to indulge in some non-reproductive sexual shenanigans. For example, masturbation is regularly enjoyed by many primates, both male and female, with orangutans rather partial to stimulating themselves with leaves and twigs.

But the most extreme sexual adventurers in the animal kingdom must be dolphins. Male bottle-nosed dolphins will have sex with practically anything that moves, including turtles, sharks and eels, while Amazon River dolphins have been observed penetrating each other's blow holes.

Many invertebrates have developed a form of internal fertilization that still utilizes spermatophores. For example, male scorpions deposit a complex spermatophore containing hooks and springs onto the ground. When a female crouches over it, the spermatophore explodes, shooting the sperm inside her.

Many spiders adopt a related strategy, in which they deposit a small blob of sperm-containing semen onto a small web. Then, using small appendages on either side of their mouth known as pedipalps, they carry this small web over to the female and manually insert the semen inside her.

Now this can be a dangerous activity because the male is usually much smaller than the female, who is prone to eating anything that comes too close to her. To avoid this fate, male spiders have developed a wide range of different strategies. A popular one is for the male to distract the female with a gift, often some form of food, while he inserts the sperm. Other organisms facing a similar danger are more imaginative: one species of fly brings the female a silk balloon to play with.

Perhaps the most fail-safe fertilization method, and the one that humans are most familiar with, is for the male to deposit sperm directly into the female. Usually, this involves the male inserting some kind of appendage, often a specialized penis, inside an opening in the female.

This is not always the case, though. Most male birds, apart from swans, ducks and ostriches, do not have penises at all; instead males and females copulate by briefly rubbing their genital openings together. The males of other organisms do have appendages for transferring sperm but don't bother inserting them into a special opening in the female. For example, small insects known as pirate bugs have a penis like a hypodermic needle and insert it straight through the female's body wall.

Some organisms seem to have developed bizarre combinations of these different fertilization methods. For example, the male of the paper nautilus, a relation of the octopus, literally fires his penis into the female, like a sort of on-heat-seeking missile.

Super-charging evolution

But why do organisms go to all this trouble to reproduce sexually when asexual reproduction is so much simpler, even for multi-celled animals. Some organisms, such as the garden snail, are hermaphrodites, able to produce both sperm and eggs, which means they are potentially able to fertilize their own eggs. But even when they have this option, many organisms only self-fertilize as a last resort; they much prefer to reproduce sexually.

The reason why most organisms on Earth reproduce sexually in one form or another is that sex super-charges evolution. Although sex can't produce new genes (only mutation can do that), it can continuously jumble those genes up to produce new combinations. These new combinations result in different patterns of genes being expressed and different traits coming to the fore, some being advantageous to the organism.

Thus, sexual reproduction provides a way for a species to respond quickly to changes in the environment. This is especially important in terms of withstanding disease and parasites, as it is more difficult for a disease-causing pathogen to wipe out a population if all the members are genetically unique. This is because at least some members of the population should have a combination of genes that makes them naturally immune. In contrast, in genetically identical populations if one member succumbs to a disease then all the members will.

So there you have it, if we want to stay healthy then we need to keep having sex, at least as a species anyway.

Spotlight: Have you finished yet?

The length of time given over to sexual intercourse, or intromission, can vary enormously between species. Stick insects, for example, are capable of having sex for weeks at a time, forcing the female to carry her constantly rutting companion around with her. Fortunately, an adult male stick insect is only about half the size of the female.

But it's not passion that brings about this impressive bout of lovemaking; it's jealousy. Stick insects are not known for their fidelity, and so a male stick insect runs the risk that a female he has just mated with will quickly mate again, reducing the chance that his sperm will fertilize the female's eggs. Unless, that is, he just keeps at it, thereby preventing any other male from getting the jump on him.

Promiscuous species often enjoy long periods of intromission, as this increases the likelihood that a male's sperm will fertilize the female's eggs before the female can find a new partner. In promiscuous mammals such as dogs, chimpanzees, mice and walruses, this seems to have led to the evolution of the baculum, or penis bone, which is exactly what it sounds like and helps the penis to stay rigid during sex.

A study in 2016 by two British anthropologists found that longer bacula were associated with more promiscuous mammals and those that tended to engage in longer bouts of lovemaking. They were also associated with mammals that have a defined mating season, when competition between males for females is particularly fierce and a good long sex session might be just the ticket.

Key ideas

- Mitosis involves a cell splitting into two, whereas meiosis involves a cell splitting into four.
- A diploid cell contains the full complement of paired chromosomes, while a haploid cell contains just a single set of chromosomes.
- Sex is the fusing of haploid cells from two parents to produce a genetically unique diploid cell.
- Male and female haploid cells are known as gametes; in animals, these gametes are sperm and eggs.
- Sexual reproduction provides a way for a species to respond quickly to changes in the environment, and is especially important for withstanding disease and parasites.

Dig deeper

Cooke, Lucy, *The Unexpected Truth about Animals: Stoned sloths, lovelorn hippos and other wild tales* (London: Black Swan, 2018).

Judson, Olivia, *Dr Tatiana's Sex Advice to All Creation: The definitive guide to the evolutionary biology of sex* (London: Vintage, 2003).

9

Man the defences

Think back to the last time you were struck down with a cold or a mild case of flu. It probably started with a just few sneezes and a cough, but before long you began to feel tired and hot, your head and muscles started to ache, and you developed a sore throat. So, you went to bed and did little but sleep for a few days, after which you began to feel better. Around a week after your first symptoms, you felt right as rain.

Congratulations, you just survived infection by one of the many viruses that cause the common cold and flu, which include rhinoviruses and coronaviruses (common cold) and the eponymous influenza viruses. And the reason you managed to survive is all thanks to your immune system, which successfully tackled and defeated the viral invader.

Saying that, almost all the unpleasant symptoms of your infection were due to your immune system rather than to the invader. Unfortunately, that is the price you have to pay for remaining alive.

The main invaders

Now, in most cases, rhinoviruses, coronaviruses and influenza viruses don't cause your immune system too much trouble, which is why you just need a few days in bed to shift a cold or a mild case of flu. But they're just the tip of the iceberg. Your body is continuously beset by a wide range of invaders, including other viruses, bacteria, fungi, protozoa and even parasitic worms, and your immune system has to deal with them all.

Why, though, does it always have to react so aggressively, irrespective of how troublesome these invaders are, especially as this reaction produces many of the unpleasant symptoms of infection? What would happen if our body adopted a more live-and-let-live attitude?

Well, the reason why these pathogens want to get inside us in the first place is because our body provides a nice, comfortable place to reproduce. Unfortunately, this reproduction usually ends up damaging our body's cells and tissues.

VIRUSES

Viruses, which are little more than collections of DNA or RNA housed within protein shells, can only reproduce inside cells, by hijacking the cells' own reproductive machinery. Once they've reproduced, they burst out of the host cells, killing them in the process. What's more, viruses can cause infected cells to fuse together and even turn into cancer cells. As well as cold and flu, viruses are responsible for diseases such as measles, chickenpox and AIDS.

BACTERIA

Bacteria can reproduce on their own, but some have a nasty habit of releasing a range of compounds that are toxic to our body's cells. These compounds, many of which are simply unwanted by-products of the bacteria's day-to-day activities, include cytotoxins, which can interfere with protein production, and lysins, which disrupt cell membranes.

Bacteria are responsible for many forms of food poisoning, such as *Salmonella* and *Listeria* infections, as well as causing tuberculosis and cholera. These disease-causing bacteria are the exception, though, because most bacteria are quite friendly (see Spotlight).

Spotlight: Bacteria are our friends

As far as most people are concerned, bacteria cause disease, as evidenced by our need for antibiotics and hand sanitizers that kill 99 per cent of all known bacteria. But bacteria don't really deserve their bad reputation. Not only could our bodies not function properly without them, but by most measures our bodies are actually more bacterial than human.

Bacteria inhabit practically every crevice, orifice and surface in and on our bodies, including our gut, mouths, scalp and skin. Our gut alone houses 100 trillion bacteria, outnumbering the cells that make up our whole body by three to one. And these aren't simply feckless squatters; they are 'beneficial bacteria' that help to digest our food, train our immune system and keep us healthy. Disturbances in the bacterial population in our gut have been linked to a whole range of diseases and disorders, including obesity, diabetes, heart disease, asthma, autism and even depression.

Colonization begins during birth, with new-born babies picking up bacteria as they pass through the mother's vaginal tract. They receive more bacteria from their mother's skin and from her milk, which not only contains lots of beneficial bacteria but nutrients to encourage their growth. As the infants start to interact with their environment and eat solid foods, their population of bacteria grows and develops.

This colonization is essential for training the immune system to distinguish between friendly bacteria and their disease-causing cousins, which is why a lot of immune-system cells live in the gut wall. A stable, diverse population of bacteria in the gut also helps to ward off invasion by pathogenic bacteria such as *Clostridium difficile*, which causes severe diarrhoea. Indeed, one effective treatment for persistent *C. difficile* infections is a faecal transplant, in which a patient is given an enema containing faeces from a healthy individual. Beneficial bacteria in the faeces colonize the patient's lower intestine, driving out the unwelcome *C. difficile*.

PROTOZOA

Protozoa, which are single-celled eukaryotic organisms, have a propensity for invading our tissues and cells, resulting in tissue damage and cell death. They are responsible for amoebic dysentery and malaria.

If left unchecked, these invaders would reproduce rapidly, eventually killing us. Indeed, if our immune system fails to defeat an invasion then that is exactly what tends to happen, unless modern drugs provide a helping hand. This is why our immune system is ever vigilant for signs of invasion and quick to destroy any unidentified foreign organisms.

Entry through the mouth or nose

First, however, the invaders have to get inside our bodies and that is far from easy. Our first line of defence is our skin, which consists of numerous layers of flat, closely packed cells and so provides an impenetrable barrier to most invaders.

As such, invaders tend to enter our body through one of our orifices, often our mouth or nose. Either we inhale them, perhaps as a result of an already infected person coughing or sneezing, or they enter within food that we consume.

BLOOD CLOTTING

This is not a risk-free entry method, however, because the mucus that lines our nose and throat contains compounds able to kill many invaders, especially bacteria. These include lysozymes, which are enzymes that can dismantle bacterial cell walls, and transferrin, which prevents bacteria from utilizing certain nutrients required for growth. The acidic environment in the stomach helps to kill many of the invaders that come in with our food.

Invaders can also enter through any breaks in our skin barrier, such as cuts. To try to prevent this from happening, our blood quickly clots, forming a scab that acts as another physical barrier. As an added benefit, clotting also prevents us from losing too much blood.

Clotting occurs as a result of cell fragments in the blood known as platelets naturally congregating around any breaks in blood vessels. This congregation immediately acts as a plug to limit blood loss. Platelets also promote the formation of a long, fibrous molecule known as fibrin, which forms a three-dimensional network of fibres over the break. This network, together with the platelets and other cells trapped within it, forms the scab.

The innate immune system

As we know from personal experience, however, it takes a few minutes for blood to stop flowing, even from a fairly small cut.

During this time, invaders have an easy route into the body. Because of this potential danger, as soon as the body detects a break in its skin barrier, it mobilizes the first wave of our immune system, known as the innate immune system.

This mobilization is triggered by mast cells in the region of the broken skin, which release a range of chemicals as a result of damage to themselves or in response to the debris produced by damage to other cells. These chemicals all act to attract various immune system cells, including neutrophils and macrophages, to the site of the break.

Ordinarily, these immune system cells circulate in the blood, where they are collectively known as white blood cells (as opposed to the red blood cells that carry oxygen around the body). One of the chemicals released by mast cells is histamine, which makes blood vessel walls more permeable. This allows large numbers of neutrophils and macrophages to pass out of the blood vessels and into the tissue surrounding the break, where they hunt for any invaders that may have slipped through.

PATTERN RECOGNITION RECEPTORS (PRRS)

They are able to do this because our bodies house a range of different compounds, collectively known as pattern recognition receptors (PRRs), that recognize and bind to characteristic molecules on the surface of bacteria, as well as to viral DNA or RNA. Some of these PRRs exist independently in blood and tissue while others are found on the surface of cells, but on binding with an invader they act as beacons for neutrophils and macrophages.

On encountering a PRR-bound invader, the neutrophils and macrophages consume them. They do this by essentially absorbing them and then digesting them within special organelles called lysosomes (see Chapter 7). If you've ever seen the film *The Blob*, then you'll get the general idea. In addition, they secrete digestive enzymes into their external environment in order to break down and clear away any cellular debris.

They also stimulate the release of hormones that raise the body's temperature, which helps to suppress bacterial growth, and make us feel sleepy, in order to conserve our energy for fighting the infection.

The innate immune response is fast and effective but also fairly indiscriminate. Large numbers of neutrophils and macrophages race to the scene of any possible infection, releasing high concentrations of digestive enzymes and gobbling up any suspicious-looking cells. It's therefore not too surprising that a few healthy body cells tend to get caught in the crossfire. Indeed, the huge influx of immune system cells, together with the resultant build-up of fluid and dead cells, produces the characteristic inflammation that occurs around a cut or injury.

The adaptive immune system

But if the innate immune system is the heavy artillery, then the next wave – the adaptive immune system – is the tactical strike force.

The crack troops of the adaptive immune system are white blood cells known as lymphocytes, which arrive after the neutrophils and macrophages. Their tactical precision is down to the fact that every lymphocyte possesses a unique class of receptor on its surface that responds to just one specific molecule.

This receptor is generated by a clever genetic mechanism, in which the genes for constructing the receptor are randomly combined from a large collection of related genes. As a result, each lymphocyte possesses a different combination that produces a unique receptor. Biologists estimate that the lymphocytes in an average adult human are probably able to detect up to 1 billion different molecules.

Like neutrophils and macrophages, lymphocytes hunt for invaders. They do this by looking for molecules that bind to their unique surface receptors. Because our body quickly destroys any lymphocytes with receptors for molecules produced by its own cells, any molecule that binds with the receptors on a lymphocyte must come from an invader. Such foreign molecules, which are often proteins from bacterial cell walls or viral shells, are termed antigens.

When a specific lymphocyte detects an antigen, which can either be floating about in the body tissue or presented to it by other cells, it reproduces rapidly. Although all the copies possess the same receptor as the original lymphocyte, they mature to perform various different functions.

Some, known as B lymphocytes, start releasing numerous copies of their receptors, known as antibodies, which bind to the invaders. Just like PRRs, these antibodies mark the invader for consumption by neutrophils and macrophages. Others, known as cytotoxic T lymphocytes, start attacking both invaders and infected cells directly by releasing proteins that can puncture cell walls and membranes, as well as chemicals that induce cells to commit suicide.

Others become what are known as memory lymphocytes. After the original infection has been defeated, these memory lymphocytes continue circulating through the body, constantly on the look-out for a repeat attack by invaders sporting the same antigen.

Spotlight: Edward Jenner, 1749–1823

Memory lymphocytes are the reason why vaccines work. The idea is to inject the body with an antigen from a specific invader, such as a protein taken from a viral coat or the cell wall of a bacterium, which will stimulate an immune response but not cause an infection.

This immune response will include the production of numerous memory lymphocytes against the antigen. These will allow the immune system to respond quickly to any actual infection, defeating the invader before it can cause any problems. Vaccines are essentially a way to prime the immune system against specific diseases.

An English scientist called Edward Jenner pioneered the development of vaccines. He noticed that milkmaids who contracted cowpox did not develop the related disease smallpox, which is much more virulent and was a scourge across Europe at the time. So, he injected pus from a cowpox blister into a young boy called James Phipps, who then proved immune to smallpox.

Through his development of vaccines, Jenner has been credited with saving more lives than any other person who ever lived.

SECONDARY LYMPHOID ORGANS

The same general immune response occurs when invaders enter our body via other entry points. In this case, however, it's more difficult for the body to realize there's a problem. With a cut, the body is immediately on the defensive, but that's not the case if the invader enters via the mouth. So, the body has come up with a system whereby any suspected antigens are quickly taken to special points spread around the body, known as secondary lymphoid organs, that possess high concentrations of lymphocytes. These organs include the spleen, tonsils, appendix and lymph nodes.

Now we can start to piece together what happens when we are struck down with a cold or mild case of flu. The initial symptoms of high temperature, runny nose, sore throat and aching joints are all down to the quick, indiscriminate response of the innate immune system and the inflammation it causes. A few days later, the adaptive immune system kicks in and begins to rid the body of invaders, while the innate response dies down. As a result, we begin to feel better.

Furthermore, memory lymphocytes ensure that we never get exactly the same infection twice, by quickly recognizing and dealing with any invader they have met before. Fortunately for us, mercy is not a quality possessed by our immune system.

Spotlight: Not immune to failure

Our immune system is good but it's not infallible. For a start, people die all the time from infectious diseases, although this is more likely if their immune system isn't operating at full strength due to old age or poor diet.

In some cases, however, invaders have evolved special strategies to help them outwit the immune system. For example, certain bacteria, including those that cause tuberculosis and Legionnaires' disease, are able to survive absorption by neutrophils and macrophages. This is either because they possess a thick, enzyme-resistant outer coat, or capsule, or have ways to stop themselves from entering lysosomes.

Conversely, the immune system can become confused and decide that innocuous molecules are actually dangerous antigens. If the innocuous molecule is produced outside the body, such as pollen or food particles, the result is an allergy such as hay fever. If the innocuous molecule is produced by the body's own cells, the result is an autoimmune disease such as eczema or diabetes.

Allergies and autoimmune diseases are becoming more common in the developed world. One potential explanation is that our modern obsession with hygiene and cleanliness means that our immune systems are not being exposed to enough invaders. As a result, they are attacking the wrong targets.

Key ideas

- Our immune system defends us against infection by various disease-causing microbes, including viruses, bacteria and protozoa.
- Most bacteria don't cause disease: the 100 trillion 'beneficial bacteria' in our guts help to digest our food, train our immune system and keep us healthy.
- Suspected antigens are taken to special points spread around the body called secondary lymphoid organs, which include the spleen, tonsils, appendix and lymph nodes.
- The first wave of our immune system is known as the innate immune system and involves mast cells, neutrophils and macrophages. The second wave is known as the adaptive immune system and involves various types of lymphocyte.

Dig deeper

Crawford, Dorothy H., *Deadly Companions: How microbes shaped our history* (Oxford: Oxford University Press, 2018).

Davis, Daniel M., *The Beautiful Cure: The new science of human health* (London: Vintage, 2019).

Yong, Ed, *I Contain Multitudes: The microbes within us and a grander view of life* (London: Vintage, 2017).

10

Attack of the nerves

At a guess, reading probably comes fairly easily to you. Indeed, you may be happily reading this page while also having a cup of tea or a glass of wine, eating a biscuit and perhaps listening to music. This will all be fairly natural for you and appears to take little or no effort.

But consider all the work your brain and nervous system are actually doing at the moment. Your eyes are scanning the page, picking out the numerous contrasts between light and dark that form the letters; your hands are carefully reaching out to grab the glass, mug or biscuit and bring it to your lips; your nose and tongue are registering the complex mixture of chemicals that comprise the wine, tea or biscuit; and your ears are reacting to a continually changing stream of sound waves.

All these physical sensations are then being fed to the brain, which uses them to construct your perception of the outside world. It transforms the contrasts between light and dark into letters, and then builds those letters into comprehensible words and sentences. It transforms the chemicals detected by receptors in your tongue and nose into attractive aromas and flavours, and transforms the stream of sound waves into a pleasant melody. And it does all this simultaneously, continuously and without you really having to think about it. Quite some feat.

The neuron

Even more impressive, this feat is all down to the activity of a simple cell; or, to be more precise, to the activity of the 100 billion nerve cells, or neurons, that make up our brain. To be even more precise, it is actually the roughly 100 trillion connections between these 100 billion nerve cells that underlie all our rather impressive sensory and intellectual abilities, including the fact that we are conscious at all.

Nevertheless, your classic neuron is rather unprepossessing, looking like a kind of elongated tree or perhaps some bizarre alien creature (see Figure 10.1). It consists of a central cell body, which houses the nucleus and other conventional cellular apparatus. Extending out from the cell body are several protrusions known as dendrites, each of which can split into numerous branches. From the bottom of the cell body extends a single, much longer protrusion known as an axon, which forms branches only at its far end. Each of these branches culminates in a golf tee-shaped structure called an end bulb or foot.

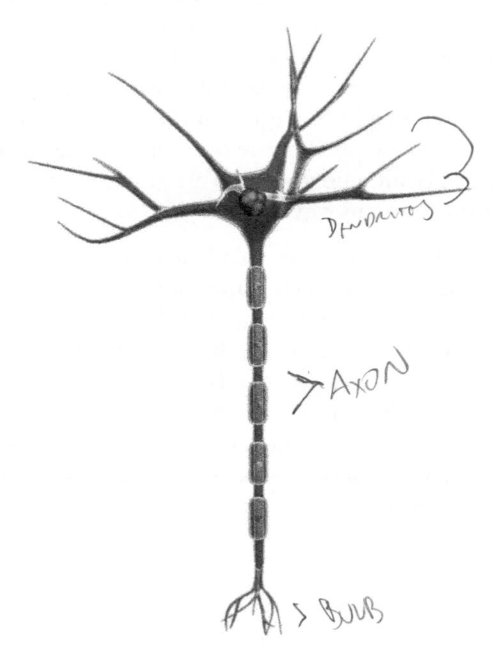

Figure 10.1 A neuron or nerve cell

Around this classic design, however, there is a great deal of variation. Human neurons range from the very small, such as many of the neurons found in the brain, to the very large, such as the single neurons that stretch around 100 cm from the bottom of the spinal cord down to the muscles of the foot.

Classic neurons, with one axon and numerous dendrites, are called multipolar and are the most common type. There are also bipolar neurons, with one axon and one dendrite and the cell body between them, and pseudo-unipolar neurons, with a single merged dendrite and axon and the cell body off to one side.

Furthermore, your classic neuron has an average of 1,000 connections with other nerve cells, but some neurons have many more. For instance, a type of neuron known as a Purkinje cell, which is found in a part of the brain called the cerebellum, has around 200,000 connections. The one common factor between all these different kinds of neuron is that each possesses at least one dendrite and one axon, even if they are sometimes merged.

CONNECTIONS BETWEEN NEURONS

This basic structure should give you a good idea of how different neurons connect with each other. Essentially, the bulbs found at the end of each neuron's axon connect with the dendrites on one or more other neurons. In most instances, however, there is no direct physical contact between the dendrites and axons, rather a small gap known as a synapse separates them.

So, what do we actually mean when we say that two neurons connect, especially as they don't actually make contact? Well, the whole point of a neuron is that it carries a signal and it is this signal that is passed between connected neurons. What is more, a neuron carries an electrical signal, just like the wires and circuits found in electronic devices such as computers. But whereas in electronic devices the electrical signal is made up of electrons, in neurons it is made up of ions (see Chapter 2).

ANIONS AND CATIONS

Positively charged cations and negatively charged anions are found throughout the body, both inside and outside of cells. Around every neuron, however, there is an unequal balance of anions and cations, with more anions inside the neuron and more cations outside of it. As a result, the inside of every neuron is negatively charged. Part of the reason for this is that neurons actively transport sodium cations out of the cell and prevent them returning by keeping special pores in their cell membranes closed.

When the neuron wants to generate a signal, known as an action potential, it suddenly opens these pores, causing the sodium cations to flood inside and transform the neuron from negatively charged to positively charged. This switch doesn't occur all over the neuron at the same time; rather it occurs in a specific section of the neuron. But the shift from negatively to positively charged in one section opens the pores in the next section and thus the action potential travels along the neuron, from the dendrites down to the axon.

After the action potential has passed, the pores close again and the neuron begins pumping the sodium cations back out, ready for the next action potential. This whole process is very quick, with the average neuron able to transmit around 200 action potentials per second.

NEUROTRANSMITTERS

When the action potential reaches the axon, it stimulates the end bulbs to release a special kind of chemical known as a neurotransmitter. Human nerve cells utilize over 50 different kinds of neurotransmitter, including small molecules such as dopamine and small peptides such as insulin.

These neurotransmitters diffuse across the synapse and stimulate receptors on the dendrite of a neighbouring neuron, producing one of two effects depending on the nature of the neurotransmitter and the receptor. Either the neurotransmitter excites the other neuron, promoting an action potential, or it inhibits it, stopping an action potential from occurring.

At any given time, a neuron is receiving signals from numerous other neurons, some of which are excitatory and some of which are inhibitory. Only when the excitatory signals outweigh the inhibitory signals does an action potential travel down the neuron to its axon.

Furthermore, neurons can form many different kinds of connections between each other. For instance, axon end bulbs don't just connect with dendrites, but can also connect with cell bodies, axons and other end bulbs. In this way, neurons can have a fine degree of control over each other's activity.

So, the human brain and nervous system comprise an extremely complex network of interacting neurons, which scientists are still a long way from fully understanding.

SENSORY RECEPTORS

But if neurons are solely concerned with transmitting action potentials to each other, how do these action potentials begin? If you traced an action potential back to its start, where would that be? Well, you would find yourself at the sensory receptors that reveal the world to us. At the rod and cone cells in the eye that detect light; at the tiny hairs in the inner ear that detect sound waves; at the taste receptors in our tongue and the olfactory receptors in our nose; and at the various pain, temperature and pressure receptors that cover our skin.

All these receptors are linked to neurons, and when they detect light or sound or a specific chemical, they stimulate an action potential in that neuron. And via the complex network of neurons, that action potential eventually makes its way to the brain.

How vision works

Take vision. Each rod and cone cell is linked to a type of neuron known as a retinal ganglion cell. There are around 126 million rod and cone cells in the human eye, with rods detecting light intensity and cones detecting colour, but only 1 million ganglion cells. As such, each ganglion cell receives signals from a set group of rods and cones.

Importantly, however, the rods and cones in each group are located in the same vicinity, which means that each ganglion cell integrates the responses from a very small patch of the retina, known as its receptive field. Furthermore, these receptive fields overlap and are also maximally responsive to changes in light intensity.

The axons of these retinal ganglion cells join together to form the optic nerve, which exits the back of each eye and makes its way into the brain, eventually arriving at a region known as the visual cortex, specifically an area termed V1. Here, the signals generated by the 1 million ganglion cells begin to be processed.

As we have already learnt, the receptive fields of ganglion cells are most responsive to changes in light intensity, which often correspond to edges, such as the edges between these letters and the white page. V1 of the visual cortex contains neurons that collectively respond to edges at all possible angles. Hence, in the first level of processing, the brain identifies all the edges present in the signal.

Other areas of the visual cortex respond to other aspects of the signal; for instance, V3 responds to depth, V4 responds to colour and V5 responds to motion. So, the signal from the ganglion cells is passed sequentially to the different areas of the visual cortex, each of which responds to a different aspect.

Spotlight: Learning from misfortune

Over the years, scientists have employed numerous methods to study the workings of the human brain. In the past, this has often involved directly stimulating different parts of the brain with electricity and noting the effects. Nowadays, scientists prefer to use non-invasive techniques such as positron emission tomography, which monitors blood flow to different parts of the brain, as a measure of brain activity.

Another way has been to study the effect of localized forms of brain damage, as a result of accidents or strokes. This has helped to reveal that the brain processes different aspects of the same sensation, such as a visual image, in different areas and so damage to these different areas produces a different set of effects.

For instance, damage to one specific area of the visual cortex causes people to lose the ability to recognize objects, both real and drawn. Damage to another area causes people to lose the ability to detect movement, such that liquid poured from a beaker appears frozen in mid-air. Damage to yet another area causes people to lose the ability to guide their hands towards objects and pick them up.

TRAPPED BEHIND OUR SENSES

In this way, the neurons in the brain transform the 1 million different action potentials from the ganglion cells into a coherent visual image. A similar process occurs with the signals coming in from the receptors in and on other areas of the body.

Although impressive, there is a slightly unnerving side to all this, because it means the world we perceive around us is actually a construct of the brain. We are trapped behind our senses, unable to perceive the world directly and reliant on our brain to provide us with an accurate representation.

Determining exactly how accurate has exercised the finest philosophical minds for thousands of years. It is also a fundamental plot point in the 1999 science-fiction film *The Matrix*, in which the hero discovers that his normal life is an illusion and that he is actually floating in a vat as a sort of human battery.

Spotlight: Taken away by Alzheimer's disease

By cutting a swathe through the 100 billion neurons that make up the human brain, Alzheimer's disease remorselessly takes away all the intellectual capabilities hitherto taken for granted, from memory to understanding to language to consciousness, eventually leading to death. A disease of old age, it is becoming increasingly common in the Western world as populations age and life expectancy increases, such that by 2050 the number of people suffering from the disease is expected to triple.

Alzheimer's disease is the most common form of dementia, characterized by the progressive loss of neurons. Its specific hallmarks are protein-based plaques and tangles in the brain, with the plaques made up of a protein called beta-amyloid and the tangles made up of a protein called tau. How and why these plaques and tangles form and what role they play in the disease is still unclear: do they cause the disease by killing neurons or are they merely a consequence of it?

Also unclear is what triggers the disease in the first place. Although some forms that strike at a comparatively young age have a strong genetic component, it has been linked with viral, bacterial and fungal infections, head injuries, smoking and pollution. Rather concerningly, there may be many different ways to trigger the wholesale destruction of neurons.

All this uncertainty has hampered the development of treatments for Alzheimer's disease, with current drugs only able to slow the progression of the disease temporarily or treat some of the symptoms, rather than the underlying condition. In any case, once the disease has started to kill off large numbers of neurons, there is no way back.

Key ideas

- The human brain contains 100 billion neurons, which form 100 trillion connections between them.
- A classic multipolar neuron comprises a central cell body, from which extends numerous dendrites and a single axon.
- Neurons connect to each other via small gaps known as synapses between a dendrite and an axon.
- When a signal known as an action potential travels along a neuron, it stimulates the axon to release a neurotransmitter into the synapse.
- By processing signals transmitted from sensory receptors around the body, our brain constructs the world that we perceive.

Dig deeper

Eagleman, David, *The Brain: The story of you* (Edinburgh: Canongate Books, 2016).

Ingram, Jay, *The End of Memory: A natural history of aging and Alzheimer's* (London: Rider, 2016).

Part Three

Earth, wind and fire

11

The ground beneath our feet

To us puny humans, the geological Earth of rocks, mountains and continents seems pretty static, merely providing the backdrop to the constantly changing kaleidoscope of life. But viewed over a timescale of hundreds of millions of years, the geological Earth becomes a great deal more dynamic.

Continents career across the surface of the Earth like dodgems, periodically smashing into each other in giant pile-ups before flying off in the opposite direction. These giant collisions form ripples in the continents that become mountain ranges, which are then gradually eroded down to nothing at all by wind and rain. Wild swings of temperature transform the surface from parched wasteland to glacier-coated snowball. Even the magnetic poles repeatedly switch positions. And this kind of breath-taking activity was taking place right from the very start of the Earth's existence.

When we left the Earth at the end of Chapter 3, or 4.5 billion years ago, it had only just reached its final mass, after agglomerating from the disc of dust and gas orbiting the Sun. Towards the end, this agglomeration process was fairly violent, with the young Earth hit by a succession of rocky bodies of various sizes, including some the size of small planets. This violent bombardment generated a lot of heat, transforming the Earth into a huge molten fireball.

The basic structure of the Earth

One outcome of this was that much of the iron within the Earth separated from the surrounding silicates, because molten iron is denser than molten silicate, and collected at the centre of the Earth, forming its core. Meanwhile, the surface of the Earth was gradually cooling, turning the outer layer of molten silicates into a solid crust.

So almost as soon as it was born, the Earth adopted the basic structure that continues to this day (see Figure 11.1). At its centre is a solid iron sphere, with a diameter of 2,400 km and a temperature of around 5,500°C, known as the inner core. Surrounding this is a 2,300-km-thick layer of liquid iron, known as the outer core, with a temperature of 4,000–5,000°C.

The great pressures at the centre of the Earth keep the inner core in its solid state, despite the temperature being more than sufficient to melt iron; in actual fact, the iron exists in several different crystalline forms. Also, iron may be the largest component (around 80 per cent), but the core probably also contains 5–10 per cent nickel (which travelled down with the iron), 7 per cent silicon, 4 per cent oxygen and 2 per cent sulphur.

THE CRUST OF THE EARTH

Surrounding the core is a 2,900-km-thick layer of hot rock known as the mantle, where the temperature increases dramatically with depth, from around 500°C near the crust to around 3,700°C near the core. On top of this floats the crust, where we and everything else live, which currently ranges in thickness from 6 km to 90 km.

Originally, however, the crust was more uniform and generally thicker, ranging from 25 km to 50 km. It consisted solely of a type of rock known as basalt.

Although the rocks making up the mantle are incredibly hot, the immense pressures within the Earth mean that (as with the inner core) they remain more-or-less solid, like a very thick gel. This

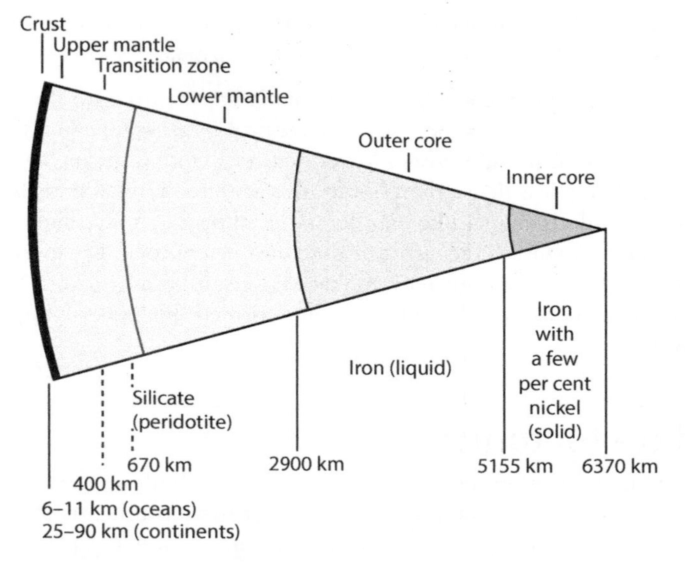

Figure 11.1 The structure of the Earth

pressure is caused by the weight of the overlying rock, meaning that it increases with depth.

Close to the surface of the Earth, the pressure can fall to a level where the rocks in the mantle are able to melt, producing magma. On the early Earth, when this magma cooled at the surface, it turned into basalt.

So, the early Earth was quickly covered in a solid crust of basalt, but that meant that the heat travelling up from the core no longer had an outlet and began to build up underneath the crust. Eventually this build-up of heat became too much, and it cracked the crust like an egg shell, allowing fresh magma to burst up through the cracks.

NEW MAGMA

This upwelling of magma forced the cracked segments of the crust apart. But because the Earth wasn't getting any larger, the far edges of those widening segments started sliding under each other, returning to the mantle. And once back in the furnace of the mantle, they started to melt.

This basalt crust was not the only thing to enter the mantle, however, because the presence of a solid crust meant that the water being delivered by comets and meteorites had started to accumulate on the surface of the Earth. So, a load of water entered the mantle along with the basalt, changing the composition of the magma being produced.

When this new magma subsequently made its way up onto the surface, either through the cracks or via proliferating volcanoes, it cooled to form rocks with a lower density than basalt, such as granite. Being less dense than basalt, this granite didn't sink as far into the underlying gel-like mantle; in the same way that a light stone doesn't sink as far into soft mud as a heavy rock. Because of this difference in buoyancy, the denser basalt became low-lying ocean floor and the lighter granite formed the higher-lying continents.

Plate tectonics

Over hundreds of millions of years, this process built up the continents that we inhabit today; it also set these continents in motion. Cracked segments of the crust are now known as plates, although these plates actually consist of the crust and the top level of the mantle, collectively known as the lithosphere. These plates continue to grow as a result of magma rising up from the mantle through the cracks, which have now become huge mountain-range-like ridges, most of which are found at the bottom of the oceans.

At the same time, the far edges of these plates continue to dive back down into the mantle, in a process known as subduction, with the denser basalt-rich ocean floor diving under the lighter continental crust. As a result, the entire plate moves, sliding over the hot mantle underneath (or, more specifically, over a layer of the mantle underneath the lithosphere called the asthenosphere). Known as plate tectonics, this process is like a conveyor belt, with the plate appearing from the mantle at one end and disappearing at the other, all driven by heat within the mantle.

DIFFERENT KINDS OF BOUNDARY

The Earth's surface is currently made up from seven major, and several minor, plates, forming a kind of irregular patchwork quilt (see Figure 11.2). These plates meet at three different kinds of boundary. The ridges where plates form are known as divergent boundaries, because two plates move off in opposite directions at each ridge. The places where two plates collide are known as convergent boundaries. At transform boundaries, two plates grind slowly past each other in opposite directions, as happens at the famous San Andreas fault in California.

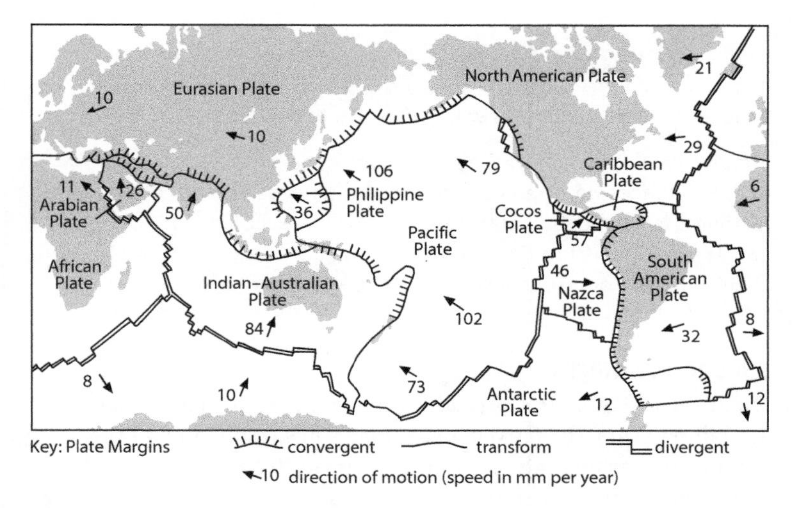

Figure 11.2 Tectonic plates

Furthermore, plates don't always dive under each other when they collide, only when dense ocean floor meets lighter continental crust. If two continental crusts collide, then they simply smack straight into each other, with the force of the collision buckling the plate ends. This is currently happening in south Asia following its collision with the Indian sub-continent around 30 million years ago, resulting in the formation and continued growth of the Himalayas.

On human timescales, plate tectonics is dreadfully slow, with continents moving at the same rate at which your toenails grow. This is one of the reasons why for a long time many scientists refused to believe it was happening (see Spotlight). On geological timescales, however, this rate of movement is sufficient to send the continents skating all over the surface of the Earth.

Spotlight: Alfred Wegener, 1880–1930

The idea that continents move and occasionally form giant supercontinents was first suggested by the German geophysicist Alfred Wegener in 1915. He was not the first person to notice that certain of the continents seemed to fit together like jigsaw pieces – for instance, South America and Africa – but he was the first to take this insight to its logical conclusion.

The reaction of his fellow geologists at the time was almost unanimously negative, especially in the United States. Although Wegener's theory, termed continental drift, explained a number of troubling findings, including why similar rocks and animals could be found in different parts of the world, it foundered on the lack of a conceivable mechanism by which continents could move.

It took the discovery of mid-ocean ridges, detailed measurements of the distances between continents and the development of plate tectonics for the idea finally to become widely accepted in the 1950s and 1960s. By this time Wegener was long gone, having frozen to death returning from a remote scientific station in Greenland.

SUPERCONTINENTS

On at least two occasions (and probably more) during the past 4.5 billion years, the movement of plates has resulted in all the continents crashing into each other to form one enormous land mass, known as a supercontinent. The first supercontinent for which there is a reasonable amount of evidence formed around 1 billion years ago and has been termed Rodinia.

It lasted for around 200 million years, before breaking up due to the build-up of heat beneath it (as happened with the first basalt crust). Around 500 million years later, a second supercontinent formed, termed Pangea, which broke up around 180 million years ago, eventually forming the distribution of continents we see today.

Spotlight: Snowball Earth

As the continents slide over the surface of the Earth, they have a major impact on our planet's climate, especially if they congregate away from the poles. If this happens, heavy rain tends to fall on the continents, eroding rocks and removing carbon dioxide from the atmosphere.

As carbon dioxide is a greenhouse gas (see Chapter 23), its removal acts to cool the Earth. This causes ice to form at the poles, which, being white, reflects heat from the Sun back into space, cooling the Earth even more. Usually, this ice covers up rock, slowing the erosion process and therefore the removal of carbon dioxide from the atmosphere. But without continents near the poles, there is no check on the cooling process, allowing ice to spread over the entire world.

There is ample rock-based evidence that the entire Earth was covered in ice on at least two occasions: around 750 million years ago and 600 million years ago. In each case, the Earth probably remained covered for around 10 million years, until the temperature started to rise again as a result of volcanoes pumping out carbon dioxide. The precise thickness and extent of this ice sheet is hotly debated, however, because there must have been some gaps for life to survive.

So, supercontinents seem to form on a 550–700 million-year cycle, which means we are on the cusp of a new one. Already, the first elements of the next supercontinent, which will probably come together in its final form in around 250 million years, are falling into place, with India crashing into Asia and Africa crashing into Europe. The Americas are still moving away from Europe, driven by the spreading caused by the mid-Atlantic ridge, which could eventually result in the Americas crashing into the far side of Asia.

THE MAGNETIC FIELD THAT SURROUNDS THE EARTH

But let's not forget the core, which despite being thousands of miles underground still has an important effect on events at the surface. For the movement of the liquid iron making up the outer core generates a magnetic field that surrounds the Earth. This field acts like a giant bar magnet, producing the north and south magnetic poles.

As well as giving us and countless magnetism-sensitive animals a way to navigate over the surface, this magnetic field provides a shield against the stream of charged particles constantly emitted by the Sun, known as the solar wind. Indeed, without this shield, life probably wouldn't have been able to get a foothold on the Earth, at least on the surface.

What is more, for reasons that scientists still do not yet fully understand, the polarity of the magnetic field repeatedly flips, with the north pole becoming the south pole and vice versa. So after the next flip a compass pointing to the magnetic north would actually be pointing towards the geographic south. Such flips happen, on average, every few hundred thousand years, although the gap between them is highly variable.

On geological timescales, therefore, the Earth is anything but static. Even on human timescales, some of the consequences of these slow geological processes can be highly dramatic.

Spotlight: Shaken to the core

More than 2,900 km beneath our feet, the Earth's core is about as inaccessible as anywhere on Earth can be. So how come we know so much about it: that it's made of iron, that there is a solid inner and a liquid outer core, that it spins? Well, much of our knowledge of the core actually comes from earthquakes, or rather the vibrations generated by powerful earthquakes.

These vibrations, or seismic waves, are so strong that they travel through the Earth, and can be detected and measured at different points on the Earth's surface by seismometers (see Chapter 12 for an explanation of how seismometers work). Crucially, these seismic waves travel through different materials at different speeds, travelling slower through denser materials but faster through more rigid materials.

What is more, there are two main types of seismic waves, known as P-waves and S-waves. P-waves are longitudinal waves (see Chapter 17), comprising alternating pulses of compression and dilation, while S-waves are shear waves, as produced by shaking a jelly. Whereas P-waves can pass through anything, S-waves are unable to pass through liquids.

So, by monitoring the seismic waves produced by earthquakes using seismometers at different points on the Earth's surface, recording what waves are detected and when, scientists have been able to determine a great deal about the Earth's internal composition. This has revealed that the inner core is solid and the outer core is liquid, with iron the most feasible main component for both.

As scientists have acquired more data from more earthquakes over the years, and developed more sophisticated ways for analysing that data and incorporating it into computer-based models, they have been able to derive even more information from seismic waves. For example, in 2015 they were able to confirm that the rising mantle plumes that cause some volcanic hot spots on the Earth's surface travel all the way up from the core.

Key ideas

- The Earth has an inner core of solid iron and an outer core of liquid iron; surrounding the core is a layer of hot rock known as the mantle, on top of which floats the crust.
- The Earth's crust is currently made up from seven major, and several minor, plates, which slide over the mantle in a process known as plate tectonics.
- The plates meet at three different kinds of boundary: divergent boundaries, convergent boundaries and transform boundaries.
- The movement of plates sometimes results in all the continents crashing into each other to form one enormous land mass, known as a supercontinent.
- The liquid iron core generates a magnetic field that surrounds the Earth, producing the north and south magnetic poles.

Dig deeper

Dartnell, Lewis, *Origins: How the Earth made us* (London: Bodley Head, 2019).

Fortey, Richard, *The Earth: An intimate history* (London: Harper Perennial, 2005).

12

Shake, rattle and roll

People in Britain are used to the quiet life, meteorologically and geologically speaking. For the most part, they don't have to put up with intense heat or cold, they don't suffer from extreme weather such as hurricanes and they don't live in fear of dangerous animals. Furthermore, the solid ground remains dependably still, and they aren't required to dodge falling lava.

But even this green and pleasant land is not entirely immune from the Earth's more aggressive tendencies. Every 100 years, Britain experiences around 120 small earthquakes, with a maximum magnitude of around 5.4, usually only causing slight structural damage and minor injuries.

Earthquakes

CHILEAN EARTHQUAKES

Other regions of the world have it much, much worse (see Figure 12.1). On 27 February 2010, a magnitude 8.8 earthquake struck the Pacific coast of Chile, knocking out electricity, water and telephone services across much of the country and damaging over 1.5 million homes.

As the epicentre of the earthquake was out to sea, it also generated a tsunami, which travelled as far as Hawaii, California, Japan and New Zealand. By the time it reached these far-off places, the tsunami had lost much of its power and the waves were fairly small. In contrast, much of the Chilean coast was battered by monster waves. The death toll was lower than initially feared, at around 500, but the damage caused by the earthquake and tsunami was valued at $31 billion.

At the time, this was the fifth largest earthquake since 1900, but was just the latest in a long line of earthquakes to hit Chile. In 1960, Chile even played host to the largest ever earthquake – magnitude 9.5. Furthermore, Chile is also littered with over 50 active volcanoes, meaning they have erupted within recorded history.

WHY CHILE?

Why should Chile be regularly beset by earthquakes and volcanic eruptions, while Britain escapes more or less scot-free? Well, it's all due to the fact that Chile lies on top of two converging tectonic plates, while Britain lies well within a plate. If you plot the epicentres of recent earthquakes and the locations of active volcanoes, you'll find that they tend to congregate at the boundaries where plates meet or are being formed (see Figures 12.1 and 12.2 and compare them with Figure 11.2).

Chile actually lies on the so-called 'ring of fire', which borders the Pacific Ocean and is one of the most geologically active areas on Earth. On the other side of the ring lies Japan, which is even more geologically cursed than Chile, being situated close to the boundaries of three plates. As a result, Japan receives 10 per cent of the world's annual release of seismic energy (vibrations within the crust), experiencing 1,000 tremors a year. Like Chile, it is also littered with more than 50 active volcanoes.

Little more than a year after the Chilean earthquake, on 11 March 2011, an even more powerful earthquake – magnitude 9.0 – struck the floor of the Pacific Ocean 300 km east of Japan, producing a 40-m-high tsunami that battered the Japanese coast. Between them,

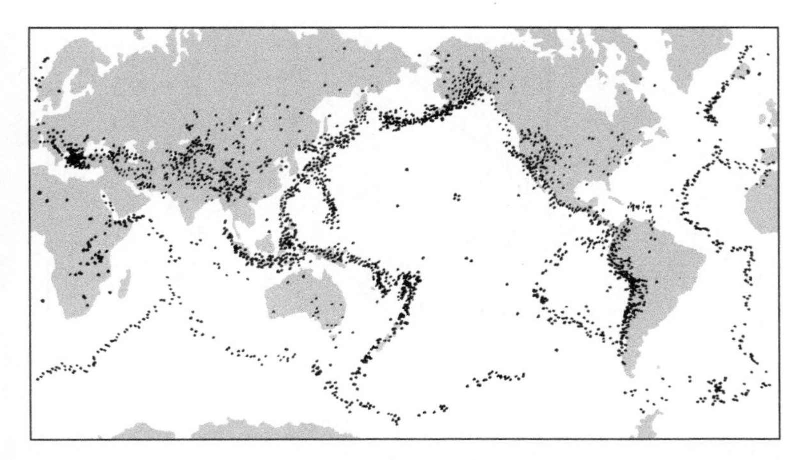

Figure 12.1 Locations of major earthquakes

the earthquake and tsunami caused 18,500 deaths and destroyed the Fukushima Dai-ichi nuclear power plant on the north-east coast of Japan's main island, resulting in the release of huge amounts of radioactive material.

Subduction

The earthquakes that struck Chile and Japan were both a by-product of subduction at a convergent boundary. All along the Pacific coast of South America, the dense ocean floor of the Nazca plate is diving under the light continental rocks of the South American plate, while east of Japan the ocean floor of the Pacific plate is diving under the Eurasian plate (see Figure 12.3).

In Chapter 10, we learned that this is an incredibly slow process, with plates moving at the same speed as your growing toenails. That doesn't mean, however, that the Nazca plate smoothly slides under the South American plate at this leisurely pace, rather it proceeds in fits and starts. The same is true for all subducting plates.

As it descends, the Nazca plate keeps on sticking to the overlying South American plate. This not only results in a gradual build-up of stress as the rest of the Nazca plate bunches up behind, but it also pulls the edge of the South American plate downward. Eventually, the stress becomes too much and the Nazca plate breaks free, lurching down towards the mantle.

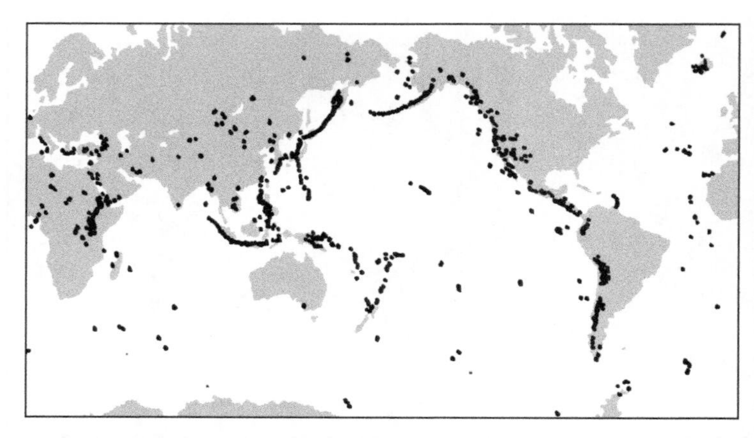

Figure 12.2 Locations of major volcanoes

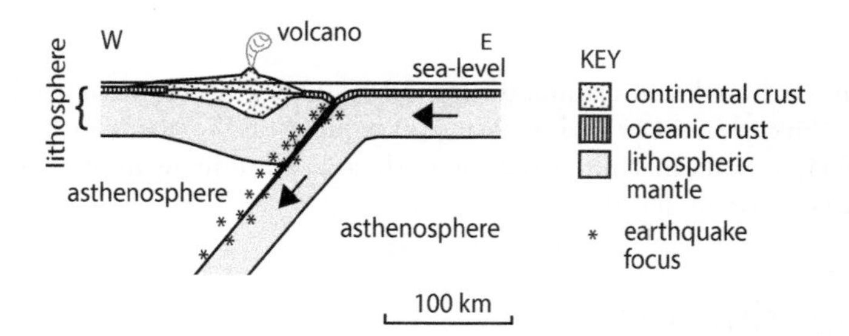

Figure 12.3 Subducting plate

In the Chilean earthquake, the subducting edge of the Nazca plate would have lurched down by around 20 metres, which is some build-up of stress when you consider that the Nazca plate only moves around 50 mm a year. This lurch downwards wouldn't have happened along the entire length of the Nazca plate edge, but only along a certain section, which would still have extended for hundreds of kilometres.

As this section of the Nazca plate broke free, the South American plate would have sprung back, like a ruler twanged on the edge of a table, causing the Earth to shake violently for around 90 seconds, producing what we call an earthquake.

Because this all took place under the Pacific Ocean, the movement of the South American plate disturbed the overlying water, creating the tsunami. As with all tsunamis, there was little evidence of it in open water, because the disturbance travelled along the bottom of the ocean. Indeed, ships often sail over tsunamis completely unaware. Only once it comes close to shore does the tsunami reveal itself.

TRANSFORM BOUNDARIES

Subduction is responsible for most of the world's largest earthquakes, including the magnitude 9.3 earthquake that struck off the coast of Sumatra on Boxing Day 2004, producing an enormous tsunami and killing over 200,000 people. But it's not the only way that plates can generate earthquakes.

The well-known San Andreas fault in California also generates its fair share of earthquakes, but they are not caused by subduction. This is because the two plates that meet at the San Andreas fault form a transform boundary rather than a convergent boundary. They rub along next to each other, as the Pacific plate moves northwards and the North American plate moves southwards.

Nevertheless, the earthquakes generated at the San Andreas fault are still caused by sticking, because once again the plates don't move past each other smoothly but proceed in fits and starts. Opposing sections of the plates stick together, causing stress to build up until eventually the plates break free and lurch forward, producing an earthquake.

But the nature of the earthquake is different. Earthquakes produced by subduction shake the ground both vertically and horizontally, whereas earthquakes produced by transform boundaries just shake the ground horizontally. Because of this horizontal movement, earthquakes at transform boundaries that are under water don't produce tsunamis.

Spotlight: Measuring the magnitude

News reports of an earthquake will always specify its magnitude: the earthquake that struck Chile on 27 February 2010 was magnitude 8.8 and the earthquake that struck off the coast of Japan on 11 March 2011 was magnitude 9.0. Everyone understands that the magnitude number corresponds to the power of the earthquake: so the Japanese earthquake was more powerful than the Chilean earthquake.

What is less well understood is that this magnitude scale, known as the Moment magnitude scale (which is an updated version of the original Richter scale), increases logarithmically rather than linearly. Each step on the scale represents an earthquake that is over 30 times more powerful (in terms of energy released) than the previous step. This means that a magnitude 9.0 earthquake is 1,000 times more powerful than a magnitude 7.0 earthquake.

An earthquake's magnitude is measured by a device known as a seismometer, which was first developed in the 1890s by a Tokyo-based British geologist called John Milne. Today, seismometers still follow Milne's basic design, consisting of three heavy weights attached to a light frame. In an earthquake, the frame moves more than the weights and this difference in movement is recorded; more powerful earthquakes produce larger differences. Each weight records movement in a different direction (two horizontal and one vertical).

Volcanoes

Transform boundaries are also not associated with volcanoes, unlike divergent and convergent boundaries (see Figure 12.2). The reason for this is that the physical conditions at divergent and convergent boundaries promote mantle melting, producing magma that is then able to rise to the surface, where it erupts as lava. At transform boundaries, however, the underlying mantle doesn't tend to melt and so volcanoes don't form.

The stereotypical image of a volcano may be a mountain-size cone with a crater at the top spraying fountains of lava all over the place, but volcanoes are very variable. Their size and shape depend on how frequently they erupt, how explosive their eruptions are and how much lava they produce.

LAVA AND PYROCLASTIC EJECTA

For as well as the characteristic lava, volcanoes also tend to produce rocky material of varying sizes, from fine ash to giant boulders weighing several tonnes, spewed over a wide area. Together, all the rocky material produced by a volcano during an eruption is known as pyroclastic ejecta.

Some volcanoes produce much more pyroclastic ejecta than lava. An example of this kind of volcano is Mount St Helens, located 200 km south of the US city of Seattle on the edge of the North American plate, which erupted spectacularly in May 1980. The force of this eruption blew away much of the volcano's side, out of which shot a lateral blast of hot gas and rocky debris that devastated 600 km^2 of surrounding terrain. This was followed by a vertical eruption of pyroclastic ejecta that reached a maximum height of 26 km.

In contrast, the Kilauea volcano in Hawaii mainly produces lava and has been erupting pretty much continuously for over 200 years, with the lava often merely seeping out rather than exploding. At Kilauea, scientists can safely obtain samples of lava by just picking them up with a spatula.

EXPLOSIVE ERUPTIONS

Whether a volcano erupts with a bang or a whimper depends on the constituency of the magma beneath it. For an eruption to be explosive, the magma needs to contain a lot of dissolved water and gases such as carbon dioxide. As the magma rises up from the mantle, it releases the dissolved gases and water (as steam) due to the decreasing pressure, in a process known as degassing, causing bubbles to form within the magma.

That's not a problem if the magma rises slowly to the surface, giving the bubbles sufficient time to escape. But if the bubbles aren't able to escape, then the magma will erupt explosively on reaching the surface. It's like shaking a can of fizzy drink. If you open it immediately, the outcome tends to be messy, but if you wait for the bubbles to disperse then there's no explosion.

Spotlight: David Johnston, 1949–80

Most scientific disciplines are not particularly hazardous, but the study of volcanoes, known as volcanology, can be. When everyone else is fleeing from an erupting volcano, volcanologists are rushing towards it. And seeing as eruptions are inherently unpredictable, volcanologists always run the danger of being caught out, even after taking all sensible precautions.

This is what happened to David Johnston, a volcanologist with the US Geological Survey and a member of the team sent to monitor the growing activity of Mount St Helens in 1980. This activity included ground tremors and a growing bulge on the volcano's northern flank.

Johnston was stationed at a supposedly safe observation point, where he was monitoring the gases being released from the volcano. But when Mount St Helens erupted on 18 May 1980, he was swept away by the unexpected lateral blast of hot gas and rocky debris. His body was never found.

Johnston's field assistant, Harry Glicken, should have been manning the observation post, but he took the day off to attend an interview. Eleven years later, Glicken was himself killed when Mount Unzen in Japan erupted.

MID-PLATE ERUPTIONS

Now, although volcanoes and earthquakes do tend to congregate around divergent and convergent boundaries, they aren't found exclusively at these locations. Even a country that is well within a plate, like Britain, can experience earthquakes. And the Hawaiian Islands, which are little more than a collection of large volcanoes whose summits poke above the waves, are located right in the middle of the Pacific plate.

Scientists still have some trouble explaining these little anomalies. They think that volcanoes in the middle of plates are caused by hot, buoyant portions of the mantle, known as mantle plumes, rising from deep within the Earth. When a plume reaches the underside of the crust, it forms a hot spot that eventually burns through, allowing magma to escape.

Recent research has indicated, however, that not all mid-plate volcanoes form in this way; instead, some may be caused by magma escaping through small cracks and faults in the Earth's crust. The movement of similar small cracks and faults in the crust also probably explains mid-plate earthquakes, although this local movement may be triggered by events taking place at distant plate boundaries.

So nowhere is truly safe from the Earth's aggressive tendencies. As well as experiencing the occasional earthquake, Britain also plays host to a few volcanoes, although none have erupted for millions of years and they are all considered extinct. This is no guarantee, however: on average, one supposedly extinct volcano erupts every five years.

Spotlight: Predicting the unexpected

Animals supposedly have no problem predicting imminent earthquakes and volcanic eruptions. There are loads of stories of animals fleeing or behaving strangely just before an earthquake strikes or a volcano erupts, although these stories are generally taken with a pinch of salt. Scientists, on the other hand, generally have much more trouble making accurate predictions.

Volcanic eruptions offer more scope for making predictions than earthquakes, because volcanoes provide detectable signs that something's about to happen. These signs include clusters of small tremors and changes in the ground level around the volcano, bulges in the side of the volcano and variations in the gases emitted by the volcano.

All these signs are caused by magma rising up from underground, and can be detected by seismometers and various other sensors around the volcano, as well as by visual observation. Ground level changes and gas emissions can even be monitored from space by satellites. The problem is that, although these signs do tend to precede a volcanic eruption, they also often appear without a subsequent eruption, making a false alarm very likely. In any case, the vast majority of volcanoes around the world aren't regularly monitored.

The plate boundaries that produce earthquakes are much harder to monitor, especially as many are at the bottom of the ocean. This means that scientists can't really predict earthquakes in advance, apart from highlighting regions of a plate that haven't slipped recently and so are due an earthquake at some point in the future.

What they can do, though, is detect an earthquake as quickly as possible with seismometers when one occurs, so they can warn people further away from the epicentre before the seismic waves have reached them. This may provide only a few minutes' warning, but can still be sufficient for people to take shelter or reach higher ground if the earthquake has triggered a tsunami.

Or perhaps they should just watch the animals. A study in 2015 found that the number of animals photographed by a series of hidden, motion-sensitive cameras in a national park in Peru dropped dramatically in the days leading up to a magnitude 7.0 earthquake.

Key ideas

- Earthquakes and active volcanoes tend to congregate at the boundaries where plates meet or are being formed.
- The so-called 'ring of fire', which borders the Pacific Ocean, is one of the most geologically active areas on Earth.
- Subduction at a convergent boundary, where one plate slides under another, is responsible for most of the world's largest earthquakes.
- As well as lava, volcanoes tend to produce rocky material of varying sizes, from fine ash to giant boulders, known as pyroclastic ejecta.
- For reasons that are not entirely understood, volcanoes and earthquakes can also occur in the middle of plates.

Dig deeper

McGuire, Bill, *Waking the Giant: How a changing climate triggers earthquakes, tsunamis, and volcanoes* (Oxford: Oxford University Press, 2013).

Rothery, David, *Volcanoes, Earthquakes and Tsunamis: A complete introduction* (London: Hodder Headline, 2015).

13

Earth rocks

In films that involve time travel, there's almost always a bit where one of the characters travels back in time to revisit an earlier scene, allowing the viewer to watch that scene unfold from a different perspective. We are now about to do something similar, by going back to the mantle.

In Chapter 11, we learned that the mantle is a thick layer of hot rock. But it is not rock in the sense that most people know it. Scientists think that the majority of the mantle, which remember is 2,900 km thick, is made up of just one kind of rock, known as peridotite.

Now, you've probably never heard of peridotite; instead, when you think of rocks you probably think of such well-known rocks as granite, slate and sandstone. Peridotite is the ancient ancestor of all these better-known rocks, transforming into them over the course of millions of years. But how does it do that?

Silicate groups

To answer this question, we must define what we mean by rocks. A rock is a collection of various naturally occurring crystalline structures known as minerals. Peridotite, for example, comprises a mixture of the minerals olivine and pyroxene.

The central component of both these minerals is a combination of silicon and oxygen called silica (SiO_2), arranged to form a building block known as the silicate group (SiO_4). This explains why the vast majority of the Earth outside the core, including 90 per cent of the crust, is made up of so-called silicate rocks. And this is just as it should be, seeing as silicate grains comprised a large proportion of the disc of dust and gas that gave birth to the Earth (see Chapter 3).

OLIVINE AND PYROXENE

In both olivine and pyroxene, silicate groups are joined by various metals: iron and magnesium are present in olivine, while iron, magnesium and calcium are present in pyroxene. Despite these similarities, however, olivine and pyroxene have very different structures.

In olivine, silicate groups are present as independent entities. This is because, individually, silicate groups exist as negatively charged anions, meaning they repel each other and never actually come into contact. They are held together by positively charged magnesium and iron cations, which insinuate themselves between the silicate anions to form the regular, crystal structure of olivine.

By contrast, in pyroxene the silicate groups are linked together in long chains, with each individual group sharing an oxygen atom with its two nearest neighbours. Iron, magnesium and calcium atoms then insinuate themselves between these chains. This causes the chains to fold up and bind with each other, like some kind of geometric office toy, forming the regular, crystal structure of pyroxene.

Spotlight: All that glitters

Minerals are not just the basis for ugly old rocks. Being crystalline, many of the minerals formed at the high pressures and temperatures found under the Earth are highly attractive, being prized as precious or semi-precious stones.

For example, silicon and oxygen don't always combine with metal ions to produce rocks. As silica, they can also keep to themselves, forming the basis for perhaps the best-known natural crystalline substance – quartz.

Although quartz in itself is not terribly valuable, the presence of impurities such as iron can turn it an attractive shade of violet, in which case it becomes amethyst.

Rocks can also house useful deposits of metals, in which case they are known as ores. The size of these deposits varies widely for different metals, as does the economics of extracting them. Iron, which is cheap and plentiful, is not worth extracting from ores that contain less than 50 per cent iron. For aluminium, the figure is 30 per cent; for lead, 5 per cent; for copper, 1 per cent; and for silver, 0.01 per cent. For gold and platinum, the scale changes to parts per million (1 gram of gold for every 1 metric tonne of rock).

Variation within rocks

In peridotite, the crystals of olivine and pyroxene come together to form a strong, interlocked network. But in fact, all rocks, including peridotite, are highly variable, meaning that even the same rocks are not identical.

For a start, there is a flexibility in the chemical make-up of minerals that isn't found in other substances. An amino acid such as glycine will always contain the same number of carbon, oxygen, hydrogen and nitrogen atoms in the same configuration.

In olivine, however, because it doesn't matter whether the silicate anions are held together by iron cations or magnesium cations, the ratio of magnesium to iron is never fixed. Some olivine will contain more magnesium than iron and vice versa. So whereas the chemical formula of glycine is NH_2CH_2COOH, the chemical formula of olivine is written as $(Mg, Fe)_2SiO_4$, because the iron and magnesium are interchangeable. Saying that, due to natural differences in abundance, the olivine in the mantle usually contains more magnesium than iron.

On top of this, the proportion of olivine and pyroxene in peridotite can also vary, as can the size of the individual mineral crystals. This means that although the vast majority of the mantle consists of peridotite, the precise nature of that peridotite can differ quite a bit. This is fortunate, as it helps to explain how peridotite can transform into all the many different types of silicate rock that make up the vast majority of the crust.

PARTIAL MELTING OF THE MANTLE

As we saw in Chapter 11, the hot rock of the mantle, which we now know is mainly peridotite, is kept more or less solid by the immense pressures within the Earth, which increase with greater depth. So, at shallower depths, such as where the mantle meets the crust, the pressure isn't so great and the peridotite can begin to melt.

Because peridotite is made up of a mixture of two different minerals with varying chemical compositions, this melting doesn't happen all at once, as with ice. Instead, different portions of the peridotite melt at different temperatures and pressures, depending on their chemical composition. This is known as partial melting.

One of the outcomes of this is that the magma produced by partial melting of the mantle has a slightly different chemical composition from the original peridotite. In particular, it results in an increased silica content: peridotite contains around 45 per cent silica, whereas the magma generated from it by partial melting contains around 49 per cent silica.

Basalt and granite

On rising to the surface at ridges, this magma cools to produce rocks collectively known as basalt, which formed the basis of the Earth's early crust and now gives rise to ocean floors. Because of the change in chemical composition caused by partial melting, basalt doesn't just consist of olivine and pyroxene, but also includes other minerals such as feldspar.

Eventually, this basalt returns to the mantle via subduction, together with ocean water, where it partially melts again, producing magma with an even higher silica content (52–66 per cent). When this magma rises to the surface, it cools to produce rocks collectively known as andesites, which are less dense than basalt and consist of a different collection of minerals.

Because they're less dense, these rocks ride higher in the mantle, forming continental crust. Partial melting can also produce rocks with even higher silicate contents (above 66 per cent), which are collectively known as granite and also form continental crust.

Basalt, andesites and granite are all examples of igneous rocks, because they form from cooling magma. There are two other basic rock types – metamorphic and sedimentary – both of which are ultimately derived from these igneous rocks.

Metamorphic rocks

Metamorphic rocks are produced when existing rocks are exposed to extended periods of high pressure or high temperature. This can occur as a result of magma rising directly through faults and cracks in the crust, in which case existing rocks are exposed to heat from the magma. It can also occur when rocks at the surface are buried by subsequent deposits of lava or sediment or when they're dragged to lower depths during subduction, increasing the pressure on them.

Although not sufficient to induce partial melting, these temperatures and pressures are able to change the nature of the minerals within the rock. Such recrystallization without melting is known as metamorphism and can produce a range of new types of rock, including such well-known examples as slate and marble.

Still, as far as the rocks are concerned, it's only when they reach the surface that things get hazardous. The minerals that make up rocks are only stable at the temperatures and pressures at which they first crystallize. Below these temperatures and pressures, such as at the surface of the Earth, they are unstable and start to chemically degrade.

Water can help this process along, especially if it contains dissolved carbon dioxide from the atmosphere. This makes the water slightly acidic, allowing it to dissolve the rock. This process tends to transform the solid rock into clay minerals such as kaolinite.

Added to this chemical degradation are the damaging physical effects of weathering. The repeated freezing and thawing of water can crack rocks and knock flakes off them, because water expands when it turns to ice. Cracks and flakes also result from the rock expanding during the heat of the day and then contracting at night. Even living organisms have a go: plants knock off bits of rock with their roots while some microbes feast on rocks (see Spotlight).

Spotlight: Rocks on the menu

Like us, lots of microbes obtain their energy by consuming and breaking down organic nutrients, but there are a few microbes that like nothing better than tucking into some good, hard rocks. These rock-eating microbes come in two main types: one type breaks down iron-containing minerals in rocks, while the other type breaks down sulphide-containing minerals.

These two types are generally found together, as the sulphide type requires a bit of assistance from the iron type to enjoy its food. When breaking down iron-containing minerals, the iron type produces a form of iron that readily reacts with sulphide-containing minerals, converting them into simpler sulphide-containing compounds that the sulphide type is then able to break down. These two types are usually joined on the rocks by more conventional microbes that feed on the organic waste produced by the rock-eating microbes.

One handy consequence of this rock eating is that it tends to release many of the metals within the rocks. Some metals, such as copper, nickel and zinc, are present in the rock as sulphides, while others such as gold are trapped within the sulphide- and iron-containing minerals. In either case, these metals are released when the microbes break down the minerals.

Because of this, rock-eating microbes are increasingly being used by the mining industry to extract these metals from ores, especially low-grade ores that contain the metals at concentrations of less than 0.5 per cent. There are several ways of doing this, but they all involve crushing the ore and then washing it with a solution of sulphuric acid containing a mixture of different rock-eating microbes.

Not only does this process use much less energy than the intense heating, or smelting, that extracts metals from higher-grade ores, but it also generates far less toxic waste. Indeed, it can be used as a cost-effective way to clean up the waste left over from old mining operations.

Sedimentary rocks

The end result is that all rocks are gradually eroded away, transformed into countless tiny particles that are either washed away by rivers or blown away by the wind. At some point, the particles, which are usually either sand or clay minerals, will settle. If transported by rivers, they will form a layer of sediment on the bottom of the river, often at the point where the river meets the sea; if transported by wind, they tend to collect as sand dunes.

Over millions of years, these layers of sediment build up, squashing and heating the layers underneath. This instigates various chemical and physical changes in the layers, most importantly the forcing out of any water, which eventually transform the soft sediment into hard rock. Examples of these sedimentary rocks include the self-explanatory sandstone and claystone.

Occasionally, the remains of living organisms become incorporated in these sedimentary layers. If those organisms possess shells made of calcium carbonate, as do many species of microscopic plankton, then limestone is the result, making it the only type of rock not ultimately derived from the mantle. If the remains of plants and algae become incorporated in the layers, being buried before they have a chance to decompose, then over millions of years the heat and pressure transform them into deposits of oil, gas and coal.

The rock cycle

So, the Earth's rocks are engaged in a constant process of recycling. Magma emerging from the mantle cools to form igneous rocks, which eventually become sedimentary rocks or metamorphic rocks, both of which can then repeatedly transform into each other (turning shells into limestone along the way). Eventually, some of these rocks may make their way back to the mantle via subduction, where the whole process begins again.

This is known as the rock cycle, but it could also be called a massive case of déjà vu.

Key ideas

- Peridotite is the ancient ancestor of better-known rocks such as granite, slate and sandstone, transforming into them over the course of millions of years.
- The Earth's crust is made up of three main types of rock: igneous, metamorphic and sedimentary.
- Limestone is produced by ancient organisms with shells made of calcium carbonate, making it the only type of rock not ultimately derived from the mantle.
- If the remains of plants and algae become incorporated in the layers of sediment, then over millions of years the heat and pressure transform them into deposits of oil, gas and coal.

Dig deeper

Rothery, David, *Geology: A complete introduction* (London: Hodder Headline, 2015).

Zalasiewicz, Jan, *The Planet in a Pebble: A journey into Earth's deep history* (Oxford: Oxford University Press, 2012).

14

Wet and windy

It may be vast, comprising 320 million cubic miles of water and covering 70 per cent of the surface of the Earth, but what is the ocean for? Well, it obviously provides us with a great deal, including food, fun, transport and even artistic inspiration, but that's not what we're talking about. If the ocean can be said to have a purpose, then its purpose is to move heat around; in particular, to help move heat from the equator to the poles.

Without this movement of heat, many forms of life would be restricted to a thin band running around the middle of the Earth. Beyond this, conditions would be too cold; indeed, without the circulation of heat by the ocean, much of the Earth would probably be covered in ice.

Take London, which is as far from the equator as Calgary in Canada, where temperatures regularly fall below –10°C in the depths of winter. In contrast, average winter temperatures in London usually stay above freezing.

Britain's comparatively mild temperature given its position is all down to the Gulf Stream and the North Atlantic Current (NAC), ocean currents that between them bring warm water over from the Caribbean on the other side of the Atlantic. Without them, Britain, together with much of western Europe, would look and feel very different.

Convection currents

The ocean is able to transport heat around the world because of convection. When water heats up it expands and becomes less dense, rising above cooler, denser water. This process sets in motion convection currents, which can be seen on a small scale when heating a glass beaker full of water. As it heats, water at the bottom of the beaker will rise to the top, displacing cooler water, which descends to the bottom to be heated. In this way, heat is circulated throughout the entire beaker.

A similar process operates in the ocean, where water is heated at the equator and cooled at the poles. In the ocean, however, the density of the water is also affected by how salty it is (or its salinity), for seawater actually differs quite a bit in salinity, depending on factors such as heat and water flow. On average, seawater contains 35 g of salt per 1,000 g of water, but this figure is lower in the Baltic Sea and Arctic Ocean, where it's cold, and higher in the Mediterranean Sea and Red Sea, where the water is confined. So, water is denser when it's cold or salty, and less dense when it's warm or fresh.

THERMOHALINE CIRCULATION

In the ocean, convection is also driven by the cold rather than the heat, in a process known as the thermohaline circulation. Let's look at the northern hemisphere first. The water brought over by the Gulf Stream and the NAC is not just warm but also fairly salty and isn't just delivered to the coasts of western Europe. In the North Atlantic, the NAC splits, with some of the water going to western Europe and some travelling up to the Arctic Ocean around Iceland and Greenland.

Here, freezing wind coursing off the Arctic ice cap saps the water of its heat, transforming salty, warm water into dense, salty, cold water, which sinks to the bottom of the ocean. Imagine a vast undersea waterfall, cascading off the coast of Greenland. This constant sinking of water drives a deep-water current that extends all the way across the Earth to the southern hemisphere.

As it travels southwards, the water gradually heats up and loses its high concentration of salt by diffusion (the propensity for particles to move from a higher concentration to a lower concentration). And as it becomes warmer and fresher, it also becomes less dense and begins rising very slowly back to the surface. On finally reaching the surface, it puts its fate into the hands of a different kind of current.

We now need to turn our gaze upwards, from the ocean to the atmosphere. In the same way that water expands and rises when heated, so does air. The Sun beating down at the equator heats the air, causing it to rise. At its summit, this tower of air splits into two, flowing towards each pole. As this flow of air cools, it drops to the Earth and splits again, travelling as a surface flow back to the equator or onward to the pole. Eventually, the poleward surface flow meets cold air coming from the pole, causing it to rise back up again and then split at high altitude, to flow either back toward the equator or towards the pole.

Circulation cells

The upshot of all this is that both the northern and southern hemispheres are embraced by three tubular air flows, known as circulation cells, which between them extend from the equator to the poles. In each cell, the air moves in a broadly circular trajectory, but in opposite directions from its neighbours. In the cells near the equator and the poles, the air moves away from the equator at altitude (and towards the poles) and towards the equator at the surface. In each of the middle cells, however, the air moves towards the equator at altitude and away from the equator at the surface. These cells are like closely meshed atmospheric cogs, with the movement of one generating opposite movement in its neighbours.

Like the ocean, these cells transport heat from the equator to the poles, although not quite as effectively (the top 2.5 m of the ocean holds as much heat as the entire atmosphere). The surface air flows generated by these cells are also better known as wind.

The direction of prevailing winds

From this explanation, you would think that these prevailing winds should blow directly north or south, depending where you are on the surface. But that's not the case. Instead, in the northern hemisphere, the prevailing winds tend to blow from the northeast in the tropics and near the pole and from the southwest at points in-between, known as mid-latitudes. In the southern hemisphere, the situation is reversed, with the prevailing winds blowing from the southeast in the tropics and near the pole and from the northwest at mid-latitudes.

THE CORIOLIS FORCE

The reason for this is that the Earth is rotating, with the equator rotating fastest and the poles rotating slowest. So, as the winds flow north or south, they end up circling the Earth faster than the ground beneath, causing them to blow at an angle. It's like a child in the middle of a roundabout spinning clockwise who throws a ball to a child at the edge. To someone standing a little way from the roundabout, the ball will travel in a straight line; but to the child at the edge, the ball will appear to slide off to the left. This is known as the Coriolis force.

Now, wind doesn't just blow hats off and keep kites flying, it also moves surface water about, with the prevailing winds generated by circulation cells setting in motion permanent currents on the surface of the ocean. Unlike the deep currents of the thermohaline circulation, which resemble rivers in the abyss, these currents are broadly circular. Again, this is because of the Coriolis force, which acts on both the prevailing winds and the resulting surface currents.

Circular currents in the ocean

The end result is that all over the ocean the surface water is travelling in huge circular currents known as gyres. For example, there are subtropical gyres that extend right across the middle of the Atlantic and Pacific oceans. The presence of land can interfere with these wind-driven gyres, setting in motion different current patterns. This happens at the South Pole, where a strong surface current known as the Antarctic Circumpolar Current continually circles around Antarctica.

Water pushed down from the Arctic Ocean can surface at any point along the way, depending on conditions. For instance, the

Mediterranean Sea is continually sending out spinning discs of salty water that can pick up the cold water travelling down from the Arctic and carry it over to the other side of the Atlantic. If the water has an uneventful journey, however, it will travel all the way to the Antarctic before surfacing.

Here, it will either get caught up in the Antarctic Circumpolar Current and be spun round Antarctica a few times, before peeling off into the Pacific Ocean or back into the Atlantic. Or it will be cooled by frozen air coming off Antarctica and sink to the bottom of the ocean again, before being pushed towards the Atlantic, Pacific or Indian oceans, where it will gradually rise to be caught up in the various gyres found in these oceans.

Eventually, the water will make its way over to the Caribbean, where it will then be dragged back along the Gulf Stream and the NAC to the frigid waters of Greenland and Iceland to begin its journey again. In total, this circumnavigation of the Earth will probably have taken over 1,000 years.

Cold, dense water sinks to the bottom of the ocean at just two places: in the polar waters of the North Atlantic and the Weddell Sea, near Antarctica. Only here, does water become cold and dense enough to sink all the way to the bottom of the ocean, dragging warmer water away from the tropics as it does so. This simple movement of warm and cold water drives the entire, complex ocean circulation system, aided and abetted by the surface currents generated by the prevailing winds.

Only recently, by taking advantage of satellite imaging and robotic buoys that sink to the necessary depths to follow deep currents, as well as marine drones that can sail the seas entirely autonomously (see Spotlight), have scientists begun to understand the full complexity of the ocean circulation system. This includes the realization that movement takes place at all scales. For example, the ocean also forms thousands of eddies – swirling water circulation systems that transport heat and salt over distances of 50–200 km – which usually only last for a few years.

Spotlight: Droning on about the ocean

Drones aren't just found flitting about the sky; increasingly they can also be found plumbing the ocean depths. But, unlike their airborne cousins, such ocean-going drones can operate with minimal human involvement. This makes them ideal for collecting a whole suite of data about ocean water, including its density, pressure, temperature and salinity, over large areas and long periods. They are also used to track pollution, watch volcanoes, measure icebergs and monitor the breaking up of ice shelfs.

These drones are essentially small submarines, powered by a battery and controlled by a computer that navigates via the Global Positioning System (GPS). In some drones, the battery is used to turn a propeller that can move the drone at speeds of around 7 kilometres per hour (kph), giving them a range of a few hundred kilometres. Other drones, known as gliders, take advantage of a less conventional means of propulsion.

In gliders, oil is moved to and from an external bladder, inflating or deflating it. This changes the glider's buoyancy, causing it to ascend or descend. Fins on the glider then convert some of this vertical motion into lateral motion, pushing it forward at a speed of around 1 kph. While slower than a propeller, this means of propulsion is much more energy efficient, allowing the glider to operate for much longer and travel much further on a single battery charge.

Hundreds of drones and gliders are now traversing the oceans, periodically rising to the surface to transmit the data they have collected via satellite and take GPS bearings to determine their position. While safe underwater from stormy weather and pounding waves, they do sometimes encounter some of the fiercer denizens of the deep, which can leave tooth marks on their outer casings.

POLLUTION IN THE OCEAN

Scientists have also mapped out the ocean circulation system by tracking the water-borne distribution of long-lasting pollutants such as the infamous, ozone-destroying CFCs (chlorofluorocarbons). A more obvious distribution of pollution can be found in the middle of the Pacific, where the North Pacific Gyre picks up much of the plastic rubbish thrown into the Pacific and dumps it into the relatively calm waters at its centre. This process is covering an area twice the size of Texas with increasing amounts of debris.

The ocean may be a source of food, fun, transport and artistic inspiration, as well as being the primary distributor of heat around the world, but increasingly we are also turning it into one, huge rubbish dump.

Spotlight: Time and tide

Tides are one of the defining features of the oceans. Twice a day, all around the world, the sea will progress a short way on to land before receding again, producing high and low tides. This is all down to the interaction between the Earth and the Moon, and to a lesser extent the Sun.

Tides are generated both by the gravitational force between the Earth and the Moon, and by the centrifugal force (which is the force that tries to eject you from a spinning roundabout) generated as the Moon orbits the Earth. The gravitational force between two orbiting bodies lessens with distance, while the centrifugal force increases the further you are away from the common centre of rotation for the two bodies.

As the Earth spins, the gravitational pull of the Moon is always strongest on the side of the Earth facing it, raising the ocean up on this side and generating a high tide. At the same time, on the opposite side of the Earth, facing away from the Moon, the centrifugal force is greatest, again pulling the oceans up and generating a high tide. At right angles to the Moon, the two forces cancel each other out, producing low tides.

Occasionally, the Sun also lines up with the Moon, exaggerating the low and high tides to produce so-called spring tides.

Key ideas

- The oceans comprise 320 million cubic miles of water and cover 70 per cent of the surface of the Earth.
- The oceans are able to transport heat around the world because of convection, with water heated at the equator and cooled at the poles.
- In the ocean, convection is driven by the cold rather than the heat, in a process known as the thermohaline circulation.
- The northern and southern hemispheres are embraced by three tubular air flows, known as circulation cells, which between them extend from the equator to the poles.
- The prevailing winds generated by circulation cells set in motion permanent currents on the surface of the ocean.

Dig deeper

Aldersey-Williams, Hugh, *Tide: The science and lore of the greatest force on Earth* (London: Penguin Books, 2017).

Kunzig, Robert, *Mapping the Deep: The extraordinary story of ocean science* (London: Sort of Books, 2000).

15

Stormy weather

The interaction between the ocean and the atmosphere is not all one way. As well as prevailing winds generating surface ocean currents (see Chapter 14), the surface ocean plays a major role in our weather – with clouds being the most obvious manifestation of this process.

Clouds and weather

Clouds are collections of water droplets and most of these droplets (around 80 per cent) come from the ocean (the rest come from lakes and other bodies of water). Sunlight heating the ocean causes water at the surface to evaporate, forming water vapour. As this water vapour rises into the air, it cools and condenses onto the multitudinous particles of dust, smoke and salt that fill the air, creating water droplets.

This process takes place all over the ocean, but it is intensified around the equator (where the surface waters are hottest) and in the vicinity of 'lows', which are regions of low pressure. For the atmosphere is not spread evenly over the whole Earth, rather it bunches up in some regions, forming regions of high pressure, and is more diffuse in others, forming regions of low pressure. These highs and lows are driven around the Earth by jet streams, which are narrow, globe-circling wind flows that operate in the upper atmosphere.

LOW PRESSURE SYSTEMS

Both highs and lows generate winds, but in opposite directions. Low pressure systems suck air in and up, while high pressure systems blow air down and out. Furthermore, the Coriolis force caused by the Earth's rotation gives the winds a bit of a sideways kick. In the northern hemisphere, winds flowing out of a high pressure system spin clockwise, while winds flowing into a low pressure system spin anti-clockwise (the situation is reversed in the southern hemisphere).

HIGH PRESSURE SYSTEMS

So, when a low is over the ocean, it actively sucks up the water vapour being generated at the surface, leading to the formation of clouds. In contrast, highs are more associated with clear skies, because water vapour is blown away by the force of the downdraft. As such, highs often produce periods of either very warm or very cold weather, both on land and at sea, depending on the amount of sunlight hitting the ground and the temperature of the air brought down from above.

Spotlight: You say El Niño, I say La Niña

The oceans are not just restricted to producing individual storms and hurricanes; they also have a hand in many large-scale weather systems. Perhaps the most famous of these is El Niño.

Ordinarily, water and winds circulate in a dependable fashion between each side of the south Pacific. Warm water is blown by prevailing winds to the east coasts of Australia and Indonesia, causing cold water to rise to the surface along the west coast of South America.

Every four to seven years, for reasons that are not well understood, this situation reverses: the prevailing winds weaken or even change direction, allowing warm water to spread over the whole of the Pacific Ocean and stopping the upwelling of cold water. This is known as El Niño and it usually lasts for a year or two.

El Niño tends to cause droughts in Australia and south-east Asia and heavier than normal rainfall along the west coast of North and South America. It has even been linked to unusual weather further afield, such as droughts in Africa. There is also an anti-El Niño, called La Niña, which periodically intensifies the usual flow, producing heavy rainfall in south-east Asia and drought in parts of the Americas.

STORM CLOUDS

If the water is fairly warm and the low extensive, then the resultant clouds can grow quite large, eventually building up into massive storm clouds, which can extend more than 16 km from top to bottom. After forming out to sea, these storm clouds can migrate with the low over land. But storm clouds can also form directly over land, especially where hot humid air (often blown in from the sea) meets cooler, drier air.

At the bottom of storm clouds, the temperature can be a comfortable 15°C. But because the atmosphere gets colder with altitude, right at the top of the largest storm clouds the temperature can be as low as –60°C. This temperature gradient not only intensifies the winds flowing through and around the cloud, but also leads to the formation of the rain and snow for which storm clouds are justifiably famous, or infamous.

The production of rain

Water vapour may get driven upwards by the wind, but once the water droplets formed by condensation become too heavy they start to fall back down. As they fall through the cloud, the droplets merge and combine to form even larger droplets, before eventually emerging from the cloud as rain. Each raindrop consists of thousands of individual water droplets.

Providing they contain a high enough concentration of water vapour, almost any size cloud can produce rain. For instance, light rain or drizzle can be produced by clouds that are just 300 m thick. But as larger clouds contain more water, they can obviously produce more rain: a single storm cloud is able to release well over 500 million litres of water.

The formation of ice crystals

Rain is not the only thing produced within large storm clouds, even though it may be all that is released. As we have seen, the temperature within storm clouds quickly falls below freezing above certain heights; at this point, the water vapour will condense to form ice crystals.

However, ice crystals can only form on particles with the right structure, usually particles of dust. On other particles, the water vapour will continue to condense as liquid droplets, even though the temperature is below freezing, producing what are known as supercooled water droplets. Unlike water droplets, ice crystals don't grow by crashing into each other, but rather by collecting more and more water vapour, which freezes directly onto the ice crystal.

All clouds with regions that are below freezing produce ice crystals, but these will usually only fall as snow if the temperature both inside and outside the cloud is low enough. Otherwise, the ice crystals melt to produce rain before they hit the ground. If the melted ice crystals are still very cold when they reach the ground, they may immediately freeze again, producing freezing rain and ice storms. Or they may encounter a shallow layer of very cold air near the ground that causes them to refreeze before landing, producing sleet.

SNOWFLAKES

If snow does fall, the exact shape of the snowflakes depends on the temperature at which they form – the classic six-sided flakes form above around –18°C, below this the flakes form as hexagonal plates. And, contrary to popular opinion, each snowflake does not have a unique shape, as duplicate snowflakes have been discovered.

Ice crystals can also combine with supercooled water droplets to produce larger, more amorphous snowflakes. Alternatively, some storms produce lots of supercooled water droplets, as a result of strong winds driving water vapour high up into the cloud. When these droplets rapidly freeze onto ice crystals, this tends to result in the build-up of hail stones, some of which can grow to the size of a large grapefruit.

Thunder and lightning

Ice crystals are also responsible for perhaps the most characteristic feature of storm clouds: thunder and lightning. As ice crystals within the cloud crash into each other they become electrically charged: one ice crystal will lose electrons to another, causing them to become positively and negatively charged respectively. The same process is responsible for the build-up of static electricity when we walk across a carpet.

For reasons that are not yet fully understood, these electrically charged ice crystals form three distinct layers within the cloud: a thick positively charged layer at the top of the cloud, an equally thick negatively charged layer in the middle and a thin positively charged layer at the bottom. When the negatively charged layer in the middle has built up enough charge, it sends a stream of electrons, known as a leader, down to the ground.

But this stream never actually reaches the ground; just before it hits, it attracts a stream of positively charged ions from the ground, which rise up to hit the cloud. It is this stream of positively charged ions, known as the return stroke, that we actually see as a lightning bolt.

Incredible forces are generated by this release of electrical charge: the average thunderstorm generates a trillion watts of lightning power, more than the combined output of all the electric power generators in the US. A single lightning bolt also heats up the surrounding air to around 30,000°C, causing the air to expand rapidly and producing a thunder clap.

Cyclones, hurricanes and typhoons

A less well-known characteristic of storm clouds is that they often rotate, as a result of the Coriolis force. This is actually obvious when you think about it, because storm clouds form above lows, which generate spiralling winds. Although all storms rotate, the poster-boys for rotating storms are the tropical cyclones, which are known by different names depending on where they form. In the Atlantic and north-east Pacific oceans, they're known as hurricanes; in the north-west Pacific Ocean, as typhoons; and in the southwest Pacific and Indian oceans, as cyclones.

Tropical cyclones are the 'daddy' of all storms, being defined as storms with winds above 74 mph (118 km/h). As their name suggests, they form in tropical waters (usually where the temperature is above 26°C), just north or south of the equator (they can't form on the equator because the Coriolis force doesn't act there). They form in the same way as all storms, sucking water vapour up from the sea; in fact, they usually develop out of normal tropical storms.

The factors that cause some tropical storms to turn into cyclones are not fully understood, although it appears to involve a runaway process in which the rising hot water vapour generates strong winds that suck up even more vapour. Each year, this process produces 80–90 tropical cyclones of varying strengths, which can extend for hundreds of miles. The destructive power of these storms, especially if they reach land, is well understood: in 2005, Hurricane Katrina devastated New Orleans, causing over 1,300 deaths and tens of billions of dollars' worth of damage.

Such destruction is caused by the high winds, which can reach speeds of 200 mph, heavy rain and, more importantly, something called a storm surge, which is responsible for over 90 per cent of all hurricane deaths. At the centre of a tropical cyclone, known as its eye, the air pressure is incredibly low, sucking up seawater. In addition, the intense winds blow masses of seawater ahead of the cyclone. The end result is that as the cyclone reaches land, it brings with it a huge wall of water than can be over 10 m high.

Some hurricanes also generate wildly spinning funnels of air known as tornadoes. If anything, tornadoes can be even more destructive than hurricanes, with wind speeds inside a tornado able to top 300 mph, but over a much smaller area. Tornadoes are usually under 60 m wide and travel for less than 16 km before petering out, although some have travelled for over 100 km.

Any storm can produce tornadoes, which means they are common wherever storms are common, especially Europe, northeast India and North America. The US is particularly tornado prone, experiencing around 1,000 a year, mainly in so-called 'tornado alley', stretching over Texas, Oklahoma and Kansas.

Whereas hurricanes cause widespread destruction, tornadoes, being smaller, tend to be more particular, and more capricious. They can destroy one house, while leaving its neighbour untouched; they can toss cars and buses into the sky, but also carefully carry a jar of pickles 25 miles without breaking it.

Robert FitzRoy, 1805–65

Today, weather forecasts are ubiquitous – on phones, websites and news programmes – and we rely on them to help us decide what to wear, where to go and what to do. This ubiquity is down in no small part to Robert FitzRoy, an officer in the British Royal Navy, who in the nineteenth century became the first person to try to establish an objective, scientific system for forecasting the weather.

Prior to this, FitzRoy had already enjoyed an eventful life. He had been a Member of Parliament and a Governor of New Zealand. His most important claim to fame, however, was captaining HMS *Beagle*, on which Charles Darwin travelled around South America and first formulated his theory of evolution by natural selection (see Chapter 5). Darwin and FitzRoy remained friends for many years, but their friendship foundered on Darwin's theory, which FitzRoy, as a staunch Christian, objected to.

In 1854, FitzRoy was appointed chief of a new department for dealing with the collection of weather data at sea, and quickly realized that this data could potentially be used to predict the weather. The data was initially provided by ships' captains, collected using instruments loaned to them. But Fitzroy also took advantage of the new electric telegraph to have daily weather reports transmitted from 15 land stations around Britain. This allowed him to create the first weather forecasts, a term he coined, which were published in *The Times* newspaper.

But FitzRoy's forecasts suffered from the same weakness that has bedevilled weather forecasts ever since: they weren't always accurate. This proved a particular problem when fishing trawlers and other ships didn't put out to sea because of a predicted storm that never arrived, making Fitzroy's weather forecasts highly controversial.

FitzRoy had always been prone to fits of anger and depression, and the strain caused by criticisms of his weather forecasts and being overlooked for a sought-after promotion, drove him over the edge. On Sunday, 30 April 1865, he killed himself by slitting his throat.

Key ideas

- Clouds are collections of water droplets and most of these droplets (around 80 per cent) come from the ocean (the rest come from lakes and other bodies of water).
- The atmosphere is not spread evenly over the whole Earth, rather it bunches up in some regions, forming regions of high pressure, and is more diffuse in others, forming regions of low pressure.
- The temperature gradient in a storm cloud not only intensifies the winds flowing through and around the cloud, but also leads to the formation of rain and snow.
- Ice crystals are responsible for perhaps the most characteristic feature of storm clouds: thunder and lightning.

Dig deeper

Craig, Diana, *The Weather Book: Why it happens and where it comes from* (London: Michael O'Mara Books Ltd, 2009).

Moore, Peter, *The Weather Experiment: The pioneers who sought to see the future* (London: Chatto & Windus, 2015).

Part Four

We have the technology

16

Full of energy

All our modern technological marvels would not be much use without a source of energy: cars need petrol; laptops need electricity; and bikes need someone to pedal. Biological organisms are just the same: without energy in the form of food they quickly stop working.

But what is energy? Well, actually, we've already defined it, because energy is simply the ability to perform work, whether moving a car, powering a laptop or synthesizing proteins.

Potential energy and kinetic energy

There are two main kinds of energy: potential energy and kinetic energy. Potential energy is stored energy, poised to perform work: a vase on a high shelf has potential energy due to the effect of gravity. Kinetic energy is the energy of motion, actively performing work: if the vase is knocked off the shelf then it has kinetic energy as it plummets towards the ground.

In actual fact, what happens is that the vase's potential energy is transformed into kinetic energy as it falls, because potential energy and kinetic energy are interchangeable. Kinetic energy is used to lift the vase up to the shelf, in the form of moving arm muscles, becoming potential energy once the vase is on the shelf. This potential energy is then transformed back into kinetic energy as the vase falls.

When the vase first starts to fall, its potential energy is greater than its kinetic energy. But as it falls, more and more of its potential energy turns into kinetic energy, until just before it hits the floor all its potential energy has become kinetic energy. At which point, this kinetic energy smashes the vase into pieces. Scientists may define energy as the ability to perform work, but that doesn't mean the work has to be useful.

DIFFERENT TYPES OF POTENTIAL ENERGY

There are actually various different types of potential and kinetic energy. As well as gravitational potential energy, there's elastic potential energy, which is the energy stored in a flexible object such as a spring when bent or squeezed. Chemical potential energy is the energy present in the chemical bonds that join atoms together into molecules, while nuclear potential energy is the energy possessed by the protons and neutrons in atomic nuclei. Electrical potential energy is the energy of positively or negatively charged particles in an electric field.

DIFFERENT TYPES OF KINETIC ENERGY

Kinetic energy can be split into translational kinetic energy, which is the energy of movement in straight lines, and rotational kinetic energy, which is the energy of circular movement. As with potential energy, kinetic energy applies at the smallest scales, with both atoms and electrons possessing kinetic energy.

This means that the pieces of smashed vase on the floor are not devoid of energy. They may no longer possess any gravitational potential energy or kinetic energy, but they still possess both chemical and nuclear potential energy and the electrons whirling around the nuclei of their component atoms still possess kinetic energy. This is collectively known as a material's internal energy.

Energy and forces

The different forms of potential energy are each associated with the four fundamental forces of nature introduced in Chapter 1. The reason for this is that for an object to possess energy there needs to be a force pulling it back to a position of lower energy.

This is most easily understood with gravitational potential energy, where the force of gravity is pulling the vase from the shelf (high-energy position) to the floor (low-energy position). But it also applies to chemical and electrical potential energy, which rely on the electromagnetic force, and nuclear potential energy, which relies on the strong force. Just like vases on shelves of different heights, some chemical bonds and atomic nuclei possess more potential energy than others, meaning energy can be released by rearranging those chemical bonds or the protons and neutrons in atomic nuclei.

Because energy continually alternates between kinetic and potential, it cannot be created or destroyed. All the energy present in the universe was there right at the start and will still be there at the very end. This is known as the conservation of energy and is the first law of thermodynamics, a set of laws that govern many aspects of energy.

From energy to heat

But the switch between kinetic and potential energy also involves something else, something that has far-reaching consequences. That something is heat. When the vase breaks on the floor, after all its potential energy has turned into kinetic energy, a portion of this kinetic energy is used to break some of the chemical bonds holding the vase together and the rest is released as heat.

The kinetic energy of the falling vase becomes assimilated within the internal energy of the vase and floor, breaking the vase and causing the molecules making up the vase and floor to move more rapidly. As these molecules are held in a tight embrace within a solid material they don't actually move around faster; instead, they vibrate more vigorously. In a gas, if the kinetic energy of the molecules increases, then these molecules do actually move around faster.

We experience this increase in molecular movement as an increase in temperature: substances with molecules that are moving fast or vibrating vigorously are hotter than substances with less active molecules. Molecules can transfer this kinetic energy to neighbouring molecules, by bumping into them or jostling them. Thus, heat passes along an iron bar when you heat one end of it.

But atoms also gradually lose their excess kinetic energy by releasing electromagnetic radiation (see Chapter 17), causing the host substance to cool down. What we feel as heat emanating from a hot iron bar is electromagnetic radiation.

GENERATING HEAT FROM FOSSIL FUELS

Now, this release of heat can be very useful; in fact, we rely on it to generate almost all the energy that we actually use. In 2016, around 96 per cent of our energy was generated by heat in one form or another, with the vast majority (over 80 per cent) derived from the chemical potential energy locked up in oil, coal and gas. As we saw in Chapter 13, these fossil fuels, as they are collectively known, are the remains of plants and algae that were compressed under rocks for millions of years. Now we burn them, which rearranges their chemical bonds and generates heat.

Unfortunately, because these fossil fuels are derived from living organisms, they contain lots of carbon. One of the molecules produced by the rearranging of the chemical bonds in fossil fuels is carbon dioxide, which has been blamed for global warming (see Chapter 23).

The heat produced by burning coal and gas can be used directly, such as to cook food and warm houses, but it is mainly used to generate electricity, by turning water into steam. This steam is used to drive a turbine, which, in turn, drives an electric generator.

What happens is that fast-moving jets of steam turn the blades of a rotor in a turbine. This rotational kinetic energy is then transferred to an electric generator, where magnets are rotated inside a metal coil. Because electricity and magnetism are two sides of the same coin, this movement generates electricity (see Spotlight).

Spotlight: What is electricity?

Electricity is actually the collective name for a range of related phenomena produced by the presence and flow of charged particles. What we term an electric current is produced by the movement of negatively charged electrons, but this movement is not like that of a flowing stream.

In fact, electrons move comparatively slowly, at less than 1 mm a second. This slow movement is driven, however, by an electric field that propagates at nearly the speed of light. So, electrons don't flow from power stations to our houses, but rather power stations generate a strong electric field that induces movement in the electrons already within our electrical devices. It is this induced movement that produces an electric current and powers these devices.

Because this process involves the electromagnetic force, the movement of electrons under the influence of an electric field generates a magnetic field. The converse is also the case, which is why moving a metal coil with respect to a magnetic field generates electricity.

Oil, on the other hand, is mainly used as a transportation fuel. Heat is again central, but the heat is released explosively. In a car, this explosion drives a piston up and down, with gears transferring this translational kinetic energy into rotational kinetic energy for the wheels. In a jet plane, it is used to drive a turbine, which generates thrust by expelling hot air at high speed.

BIOFUELS

Heat can also be produced by burning plants and waste material. Like coal and gas, this heat can be used directly or to produce electricity. But like oil, this heat can also be used to power vehicles, as long as the plants and waste material are first converted into liquid biofuels such as bioethanol and biodiesel.

NUCLEAR POTENTIAL ENERGY

Finally, heat can be derived from nuclear potential energy, by splitting atoms in nuclear power stations. This is known as nuclear fission and involves hitting uranium atoms with neutrons, breaking the bonds that hold together the protons and neutrons in the uranium atom nucleus. This produces new atoms, such as barium and krypton, more neutrons, which go on to split more uranium atoms, and lots of heat. Just like coal and gas, this heat is then used to turn water into steam for producing electricity.

ENERGY NOT GENERATED BY HEAT

At the moment, only around 4 per cent of the energy we use is not ultimately generated by heat. Collectively termed renewable energy, this includes the energy generated by wind, wave, tidal and hydroelectric power, which all still use turbines to generate electricity. But they use the movement of water or wind to drive the turbine, rather than steam.

Only one energy technology does not utilize either heat or turbines and that is solar power. Here, a device known as a solar, or photovoltaic, cell takes advantage of the semiconducting properties of silicon (see Chapter 18) to convert sunlight directly into electricity.

Spotlight: Storing for later

Although still dwarfed by oil, coal and gas, renewable energy technologies such as solar and wind are producing an increasing proportion of our energy. The main advantage of renewable technologies is that they don't generate carbon dioxide, and so don't contribute to global warming, but this comes at the expense of flexibility.

Most renewable energy technologies produce energy only under certain conditions, such as when it's sunny for solar and when it's breezy for wind. This makes them prone to generating too much energy when it's not needed and too little energy when it is. Hence the need for ways of storing the energy they produce for future use.

One obvious way to store the excess energy produced by renewable energy technologies is to use the same kind of rechargeable batteries that power our mobile phones and laptops, known as lithium-ion batteries. This is being done in South Australia, where the world's largest lithium-ion battery, switched on in 2017, can store up to 100 megawatts of electricity produced by a nearby wind farm. But lithium-ion batteries represent an expensive way to store large amounts of energy.

Much cheaper, and the most popular current way for storing energy, is pumped-storage hydropower. This involves using excess energy to pump water from a low reservoir to a higher one, and then letting the water flow back to the lower reservoir, spinning a turbine as it does so, when energy is required. Scientists have developed several variations of this process that work with other materials, such as pumping gas between two chambers or transporting rock-hauling railway cars between different heights.

They are also developing new battery technologies for storing large amounts of energy. These include redox flow batteries, which generate electricity by flowing two ion-containing liquids either side of a special membrane. The liquids are held in huge tanks, where they are recharged; this means the battery's storage capacity can be increased by simply increasing the size of the tanks.

Loss of energy

But heat also has a dark side, because it gradually reduces the amount of usable energy in the universe. This dark side manifests itself in the inefficiency of our current energy generation technologies. As can be imagined, burning coal to produce heat to turn water into steam to rotate a turbine to generate electricity is not a terribly efficient use of the chemical potential energy locked up in coal.

The electricity produced by a typical generator possesses at most 38 per cent of the energy used to produce the steam, meaning 62 per cent of the chemical potential energy is simply lost. Furthermore, another 5–10 per cent of the energy is lost as a result of transmitting this electricity through cables, which can get very hot. As a consequence, boiling water in an electrical kettle takes three or four times as much energy as boiling the same amount of water over a flame.

All our current energy generation technologies, including solar, lose large amounts of energy. This energy is lost as heat to the environment and once lost it can't be utilized again. We can make intense heat perform work, but diffuse heat can't be utilized for anything. This loss of energy as unusable heat is known by scientists as an increase in entropy and is the basis for the second law of thermodynamics.

And what is true of the Earth is true of the universe in general. Stars like the Sun generate monumental amounts of energy, but most of this is simply lost into space, where it does nothing more useful than slightly warming the universe. Because of this, the amount of usable energy in the universe is steadily decreasing, with potentially alarming implications for its long-term future, as well as ours (see Chapter 30).

Spotlight: $E = mc^2$

It is without doubt the most famous equation in the whole of science: adorning T-shirts and posters, as well as being the title of a hit 1986 song by the rock group Big Audio Dynamite. Worked out by the great German-born physicist Albert Einstein in 1905, $E = mc^2$ says that, at heart, all matter is simply energy.

E stands for energy, *m* stands for mass and *c* stands for the speed of light, which is multiplied by itself or squared. By showing just how much energy could theoretically be liberated from matter, the equation acted as a signpost for the development of new forms of energy production.

It directly led to the idea that atomic nuclei could be split or fused with each other to produce energy, resulting in both the atomic bomb and nuclear power. It also led to increasingly powerful particle accelerators, which not only convert matter into energy but also energy into matter.

Key ideas

- Energy is defined as the ability to perform work, whether moving a car, powering a laptop or synthesising proteins.
- There are two basic kinds of energy: potential energy, which is stored energy, and kinetic energy, which is the energy of motion.
- Energy cannot be created or destroyed; this is known as the conservation of energy and is the first law of thermodynamics.
- Substances with molecules that are moving fast or vibrating vigorously are hotter than substances with less active molecules.
- The loss of energy as unusable heat is known by scientists as an increase in entropy and is the basis for the second law of thermodynamics.

Dig deeper

Coopersmith, Jennifer, *Energy: The subtle concept* (Oxford: Oxford University Press, 2015).

Rhodes, Richard, *Energy: A human history* (New York: Simon & Schuster, 2018).

17

Coming in waves

To transfer energy over large distances, you need waves. This can be appreciated by anyone who likes to bask in the waves of light and heat emanating from the Sun, almost 150 million kilometres away.

Waves are disturbances that propagate over a distance and can be generated by the more localized movement of certain objects, from atoms to molecules to guitar strings to sections of the Earth's crust. A classic example of a wave can be seen in the ripples produced when a stone is dropped into a still pond. The ripples are known as transverse waves, because the disturbance occurs at right angles to the movement of the wave (the ripples spread out horizontally, while moving up and down vertically).

The other classic type of wave is known as a longitudinal wave, in which the disturbance moves in the same direction as the wave. Longitudinal waves can easily be produced in a spiral tube of thin wire, otherwise known as a slinky, by moving one end back and forth. This causes a region where the wires are squashed together to travel down the wire, followed by a region where the wires are more spread out. This is a longitudinal wave and is the form taken by sound waves travelling through the air, where they comprise regions of high and low pressure, and by certain seismic waves travelling through the Earth (see Chapter 11).

Wavelength, amplitude and frequency

Both transverse and longitudinal waves possess three main characteristics: wavelength, amplitude and frequency. These characteristics can most easily be explained by considering the simplest type of transverse wave, known as a sine wave (see Figure 17.1). Now a sine wave has a formal, mathematical definition, but it is essentially what most people think of as a stereotypical wave.

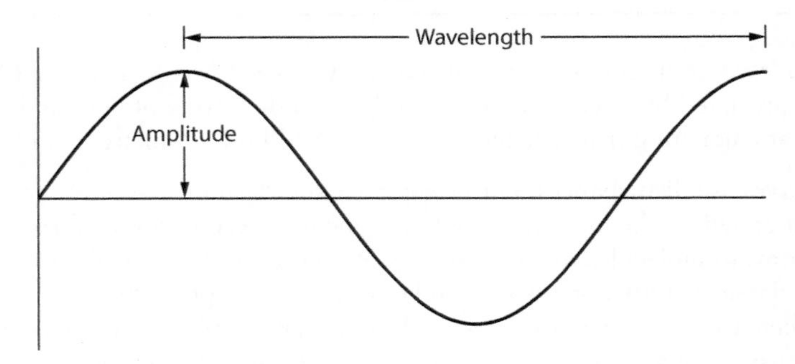

Figure 17.1 Sine wave

The wavelength is the distance covered by a single cycle of the wave, such as the distance between two consecutive peaks, measured in metres. The amplitude is the maximum disturbance from the equilibrium position: for a transverse wave, this means the height of the peak or depth of the trough in metres; for longitudinal waves, it means the maximum or minimum compression. The frequency is the number of waves passing a fixed point in a second, measured in hertz (1 hertz or Hz equals one wave, or cycle, a second).

Another important characteristic of the wave, its speed of propagation, can be calculated by simply multiplying the frequency by the wavelength. So, a wave with a frequency of 5 Hz and a wavelength of 2 m is travelling at a speed of 10 m a second.

Waves and energy

As energy is required to produce movement, whether in an atom or part of the Earth's crust, the subsequent wave also possesses energy, with the amount of energy reflected in the wave's amplitude and frequency.

Now, it's obvious that waves with larger amplitudes possess a greater amount of energy. This is demonstrated by sound waves, where the amplitude is directly related to loudness, and waves on a beach, where small waves lap your feet but large waves knock you over. Frequency is also a factor because the higher the frequency the more energy-possessing waves hit a certain point over a set period of time.

It therefore takes more energy to produce waves with higher frequencies and amplitudes, as can be seen with electromagnetic radiation.

Electromagnetic radiation

Rapidly moving sections of the Earth's crust produce earthquakes, rapidly moving guitar strings produce music and rapidly moving atoms and molecules produce electromagnetic radiation. More precisely, electromagnetic radiation is produced by the movement of the charged electrons and protons that make up atoms and molecules.

Electromagnetic radiation is a transverse wave and can adopt a huge range of different wavelengths and frequencies: from radio waves at 100 Hz with wavelengths measured in thousands of kilometres to gamma waves at 10^{20} Hz with wavelengths measured in fractions of a nanometer (see Figure 17.2). The form of electromagnetic radiation that we are most familiar with, however, is visible light, which comprises a very thin band of the electromagnetic spectrum with wavelengths between 400 nanometres (nm) and 700 nm and frequencies of 400–800 trillion Hz.

All electromagnetic radiation travels at the speed of light (300,000 km a second; although see Spotlight). As the speed of propagation of a wave is simply the wavelength multiplied by the frequency, this explains why the frequency and wavelength of electromagnetic radiation are inversely related. For all forms of electromagnetic radiation to travel at the speed of light (and nothing can travel faster than the speed of light), the frequency must increase as the wavelength decreases and vice versa.

Although electromagnetic radiation is a transverse wave, it actually consists of two parts: an electrical part and a magnetic part. Indeed, electromagnetic radiation is best thought of as oscillating electric and magnetic fields travelling through space. The two parts travel in phase (meaning that the peaks of the electrical wave match up with the peaks of the magnetic wave), but at right angles to each other (so that the magnetic wave appears as a kind of shadow to the electrical wave).

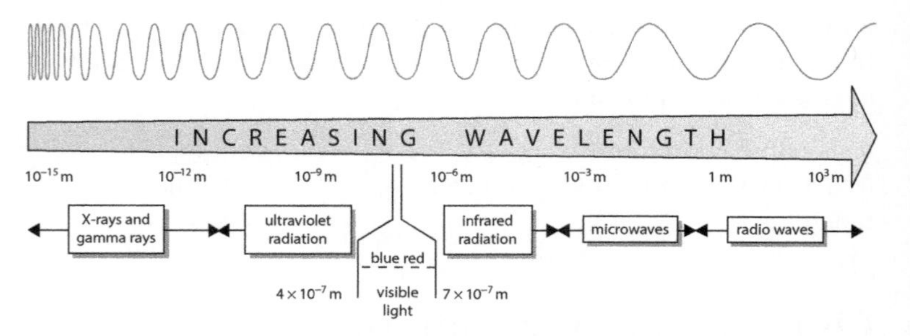

Figure 17.2 Electromagnetic spectrum

DIFFERENT FREQUENCIES

Atoms and molecules don't emit electromagnetic radiation at just a single frequency, but over a range of frequencies, with the precise range depending on how energetically they are moving. Rapidly moving or vibrating atoms and molecules (with lots of kinetic energy) emit electromagnetic radiation at higher frequencies than less rapidly vibrating atoms and molecules.

At a single frequency, electromagnetic radiation exists as a classic sine wave, but if the radiation is emitted over a range of frequencies then the individual sine waves combine to produce more complex waves. These are much less regular than the classic sine wave, with the precise shape determined by how the peaks and troughs of the individual sine waves combine. Nevertheless, because all electromagnetic radiation is simply made up of combinations of different sine waves, it can theoretically be split into its individual frequencies.

LIGHT AND COLOURS

Take visible light, which as we have seen is made up of a narrow range of frequencies. To split visible light into its component frequencies, all you need do is hold up a prism, which transforms white light into all the colours of the rainbow (see Spotlight below for an explanation of how this works). All the colours that we experience are simply different frequencies of electromagnetic radiation: from 400 trillion Hz for red light to 800 trillion Hz for violet light.

Spotlight: Rainbows in a blue sky

The reason a prism is able to separate white light into its component frequencies is because electromagnetic radiation only travels at the speed of light in the vacuum of space. Passing through any kind of substance slows it down, with solid materials slowing it the most and gas slowing it the least. This explains why a straw seems to bend in a glass of water, because light is deflected by the change in speed as it goes from the water to the air.

On top of this, the degree to which electromagnetic radiation is slowed depends on its frequency. So, when white light passes through a transparent prism, high frequencies slow more than low frequencies, causing the light to be split into all the colours of the rainbow. The same process also produces rainbows in the sky, with raindrops taking the place of the prism.

Light can also be scattered by molecules and other tiny particles, with the amount of scattering again dependent on the frequency of the light: higher frequencies are scattered more than lower frequencies. The reason the sky is blue is that gas molecules in the atmosphere scatter high frequency blue and violet light more than low frequency red light, causing blue light to appear to come from all points of the sky.

The colourful world that we see around us is entirely due to the fact that different materials absorb and reflect different wavelengths of light. So, your T-shirt appears red because it reflects red light and absorbs all other frequencies.

INFRARED FREQUENCIES

For atoms and molecules don't just emit electromagnetic radiation, they can also absorb it, with different atoms and molecules absorbing different frequencies (see Spotlight). This causes them to move more rapidly and so emit electromagnetic radiation at higher frequencies.

Spotlight: Absorbing frequencies

The tendency for specific molecules to emit or absorb electromagnetic radiation at certain frequencies provides a means for identifying them. For example, our rather astonishing ability to detect molecules in molecular clouds many, many light years away (see Chapter 2) rests on the fact that molecules tend to emit and absorb electromagnetic radiation at specific radio and infrared frequencies.

Pointing a radio or infrared telescope at a dense molecular cloud illuminated by a background star produces a read-out of electromagnetic radiation at different frequencies, known as a spectrum. Without the cloud, this spectrum would contain an equal mix of the different frequencies. But because molecules in the cloud absorb and emit electromagnetic radiation at different frequencies, it actually contains peaks and troughs.

A molecule that emits light at a specific frequency will show up as a peak in the spectrum, while a molecule that absorbs light at a specific frequency will show up as a trough. By comparing these spectra with those produced in laboratories on Earth, scientists can confidently identify many of the molecules responsible for the peaks and troughs.

As we saw in Chapter 16, atoms and molecules moving more rapidly translates into a temperature increase in the host material, whether gaseous, liquid or solid. If a material absorbs electromagnetic radiation, it will heat up, which explains why almost everything gets hot on a sunny day. It will then cool down by releasing electromagnetic radiation, usually at infrared frequencies, which are just below the frequencies of visible light (from around 400 trillion Hz to 1 trillion Hz).

RADIO WAVES

Below the infrared frequencies are microwaves (from around 300 billion Hz to 300 million Hz) and below this are the radio waves, which form the basis for all forms of long-range communication (see Spotlight, below). As we said earlier, it takes less energy to produce electromagnetic radiation at low frequencies than high frequencies. Radio waves with frequencies of a few million hertz can be produced by simply applying a varying electric current to a metal aerial. The resultant flow of electrons in the aerial generates radio waves, with the precise frequency of the waves depending on how rapidly the current is varied.

So, to produce a radio signal all you need do is use a microphone to convert sound waves, such as speech or music, into a varying electric current. This is usually accomplished by connecting something that responds to sound waves, such as a diaphragm, to an electrode, such that the movement of the diaphragm generates an electric current. This signal is then passed straight to the aerial, producing radio waves.

Radio waves with a frequency of a few thousand hertz can travel for up to 1,000 miles before losing energy and petering out. This means they can be picked up by a receiver some distance away, in the opposite of the transmission process, with the radio waves generating a small electric current in a receiving aerial. This current is then transferred to headphones or a speaker, which convert the electrical signal into sound waves.

Spotlight: Guglielmo Marconi, 1874–1937

The Italian inventor and engineer Guglielmo Marconi did not discover radio waves – that was the German physicist Heinrich Hertz, whose reward was having his name memorialized as the unit of frequency. He was also not the first to develop an instrument for transmitting and detecting radio waves, as that had already been achieved by several scientists, including Hertz and the British physicist Oliver Lodge. He was not even the first to transmit radio waves over long distances, as he achieved that feat at around the same time as the New Zealand physicist Ernest Rutherford. But he was the first to realize the potential of radio waves for long-distance communication.

Following the success of the electric telegraph, which allowed messages to be sent instantaneously over wires, including across the Atlantic, researchers were investigating various ways to achieve 'wireless telegraphy'. Various approaches had been attempted, although without much success, until Marconi realized that radio waves might provide the answer. In contrast, Rutherford and his fellow scientists were purely interested in investigating the scientific implications of radio waves.

Marconi came from a rich family and began his experiments on his father's estate, at the age of just 21, eventually managing to send radio signals over a distance of one and a half miles. He then took his work to Britain, where he was granted the world's first patent for a system of wireless telegraphy. Over the next few years, he steadily improved the technology and demonstrated radio transmission over longer and longer distances: 12 miles in 1897, across the English Channel in 1899 and then across the Atlantic Ocean, between Britain and Canada, in 1901. The modern era of global communications had begun.

Unlike Hertz, Marconi wasn't rewarded with his own unit of measurement, but he did receive many other honours, including the Nobel Prize in Physics in 1909.

CONVERTING SOUND WAVES TO RADIO WAVES

Almost all forms of long-range communication, from radio to television to mobile phones to Wi-Fi, utilize radio waves in this way. Obviously, quite a few technical hurdles have had to be overcome along the way. One of the earliest was how to convert sound waves, which have a frequency of a few hundred hertz, into radio waves with a frequency of a few million.

To do this, early radio engineers came up with a process known as modulation. The earliest form of this is known as amplitude modulation (AM), which involves altering the amplitude of a high frequency wave to match the peaks and troughs of a low frequency wave. An alternative version, known as frequency modulation (FM), does the same thing by altering the frequency of the high frequency wave and has the advantage of suffering from less interference. This is why we now have AM and FM radio.

TELEVISION

Television offered a further challenge, because now images as well as sound needed to be transmitted. To do this, engineers adopted the same basic technique, but just with a greater range of frequencies. So black and white images are sent on one set of frequencies, in the form of waves representing the brightness of hundreds of different squares, or pixels, that together form the image. Colour, sound and synchronizing pulses to match the different signals together are then all sent on separate sets of frequencies.

Increasingly, however, radio waves are no longer transmitting physical representations of sound waves or visual images, but rather just strings of ones and zeros.

Welcome to the digital age.

Key ideas

- Waves are disturbances that propagate over a distance and can be generated by the more localized movement of certain objects.
- There are two classic types of wave: transverse waves, where the disturbance occurs at right angles to the movement of the wave, and longitudinal waves, where the disturbance moves in the same direction as the wave.
- As energy is required to produce movement, the subsequent wave also possesses energy, with the amount of energy reflected in the wave's amplitude and frequency.
- Light is a form of electromagnetic radiation, with wavelengths between 400 nm and 700 nm and frequencies of 400–800 trillion Hz.
- Other forms of electromagnetic radiation include infrared, microwaves and radio waves, which form the basis for almost all forms of long-distance communication.

Dig deeper

Goldsmith, Mike, *Waves: A very short introduction* (Oxford: Oxford University Press, 2018).

Walmsley, Ian A, *Light: A very short introduction* (Oxford: Oxford University Press, 2015).

18

Information overload

Think of all the books in all the libraries of the world – from the US Library of Congress to the smallest mobile library trundling around the English countryside. Think of all the information present in those books: the vast collection of knowledge, opinion and experience expressed in countless words.

Well, all those books count for just a tiny proportion of the total amount of information currently being produced by humankind; a proportion that is falling all the time. And rather than finely crafted words, this huge mass of information consists of nothing more than strings of zeros and ones – 0s and 1s.

Sounds impossible to believe? Think of all the emails you send and receive each day, and the tweets, Instagram posts and Facebook updates; think of all the music, films, box sets and computer games you stream; think of the books, newspapers and magazines you read on your phone and the photos you take; think of all the blogs, vlogs and videos of funny cats.

All of this is information that exists, ultimately, in the simple form of strings of 0s and 1s, known as binary digits. And this digital information is growing rapidly, requiring us to continually update our terminology. In 2005, it's estimated that humankind created, copied and transferred 150 exabytes (150 billion gigabytes) of data. Eight years later, the figure was probably something like 2,000 exabytes (2 zettabytes). By 2025, it could well be 180 zettabytes.

Boole and logical reasoning

But how did we get to this point? Well, in many ways, we can trace the current deluge of information back to the insight of a young US engineer called Claude Shannon, who while studying at the Massachusetts Institute of Technology in the late 1930s became very interested in electrical switches. In particular, he became interested in the possibility of using electrical switches to carry out a form of symbolic logic known as Boolean algebra.

Devised by a British mathematician called George Boole in the mid-nineteenth century, Boolean algebra allows logical reasoning to be expressed in a mathematical form. Logical reasoning involves determining whether a specific statement is true or false based on whether certain other statements are true or false, and is thus the basis for all rational argument. To give a simple example, if the statement 'All dogs are mammals' is true and the statement 'A Labrador is a dog' is true, then the statement 'All Labradors are mammals' must also be true.

Boole showed that such logical reasoning could be recast as mathematical formulae, in which the truth or falsity of specific statements are inputs to logical rules that produce outputs that can also be true or false. The above example demonstrates an AND rule in operation: if the two input statements are true then the output statement is also true; but if one or both of the input statements are false, such as if we replace 'mammal' with 'reptile', then the output statement will also be false (All Labradors are not reptiles).

Spotlight: Alan Turing, 1912–54

Claude Shannon may have been one of the pioneers of modern computing, but there were many others as well, on both sides of the Atlantic. One of the British pioneers was the mathematician Alan Turing. In a 1936 academic article on the rather arcane mathematical subject of computational numbers, Turing first set out the theoretical design of a modern computer. This comprised a mechanical object, now known as a Turing machine, performing certain functions according to external instructions and its own internal state.

In 1950, he designed the so-called Turing test for determining whether a computer was intelligent. This involved an experimenter putting questions to an unseen human and computer; if the experimenter couldn't tell from the answers which was which, then the computer could be deemed intelligent.

During the Second World War, Turing also played an important role in breaking some of the German military codes, including the notorious Enigma codes. Tragically, he died young, committing suicide after being prosecuted for homosexuality, which was then illegal.

THE DEVELOPMENT OF THE TRANSISTOR

Because the inputs and outputs in Boolean algebra can only adopt two values, true or false, Shannon realized that they could be represented by electrical switches, which can be either on or off. This means that if you join enough of these switches together, they should be able to perform logical reasoning. It took another 10 years and the development of the transistor for this idea to really take off.

A transistor is essentially a tiny electrical switch made from a semiconducting material based on silicon or germanium. As its name suggests, a semiconductor sometimes conducts electricity very well and sometimes not at all, depending on certain conditions. By controlling those conditions, the transistor can be toggled between conducting and non-conducting states, thereby acting as a switch.

Once they had these tiny switches, scientists could join them together to perform specific logical functions, producing what are known as logic gates. So, an AND gate has two inputs, which rather than being true or false are 1 or 0, representing either the presence or absence, respectively, of an electrical current. If both of the inputs are 1 then the output is 1, whereas if one or both of the inputs are 0 then the output is 0.

The two other basic logic gates are an OR gate and a NOT gate. In an OR gate, if one or both of the inputs are 1 then the output is 1, only if both inputs are 0 is the output 0. A NOT gate has only one input and one output, which is simply the opposite of the input: so, a 1 becomes a 0, and vice versa. These patterns of inputs and outputs for each logic gate can be presented as tables, known as truth tables.

This all sounds very simple, but if several logic gates are connected together into simple circuits, such that the output of one logic gate becomes the input of another logic gate, then they can perform basic computing functions, such as arithmetic. But rather than this arithmetic being conducted in our familiar decimal system, which is based on multiples of 10, it needs to be conducted in binary, because the only numbers available are 0 and 1.

From decimals to binary

In the decimal system, numbers increase in size to the left, with each additional figure representing an increase by a factor of 10. So, the number 351 is made up of one single digit, five 10s and three hundreds; if we added another figure to the left of the 3, then that would represent thousands. It is the same with binary, but here each additional figure represents an increase by a factor of two. So, the number 1101 is made up of one single digit, one four and one eight (the zero in the second position means there is no two in this number). As such, 1101 is equivalent to the decimal number 13 (1 + 4 + 8).

Every decimal number has its binary equivalent, allowing suites of logic gates to perform addition, subtraction, multiplication and division. But that's just the tip of the iceberg, because it's not just decimal numbers that can be converted into binary form. Any kind of information can be converted into strings of 1s and 0s.

CONVERTING SOUND INTO 1S AND 0S

Take sound waves (see Chapter 16), which would seem pretty resistant to conversion into 1s and 0s, seeing as sound waves are continuous and 1s and 0s are individual digits (hence the term digital). But all you need do is essentially cut the wave into lots of thin strips and then measure the amplitude of the wave in those little strips. This will produce a string of decimal numbers that change with time and, as we have seen, decimal numbers can easily be converted into binary.

If these 1s and 0s are represented by tiny pits on a flat plastic disc (where the presence of a pit represents 0 and the absence represents 1), then you have a CD. If instead of encoding just sound waves, these pits also encode brightness and colour then you have a DVD.

USING BINARY DIGITS TO STORE DATA

Rather fortuitously, it turns out that the most efficient way to encode, store and transmit any information is as binary digits, also known as bits (see Spotlight, below), and it was Shannon who proved this. Say you want to display all the numbers from 1 to 999 in tokens, what is the minimum number of tokens you would need? If you were displaying the numbers as decimals, then you would need 29 (two sets of 0 to 9 and one set of 1 to 9).

If, on the other hand, you were displaying the numbers as binary then you would only need 20 (10 sets of 1s and 0s), because this would allow you to represent every number up to 1,024 (2 multiplied by itself 10 times, or 2^{10}). The same is true of all information: there is no more economical way to encode it than as binary digits.

Spotlight: Bits and bytes

The term 'bit' is a contraction of 'binary digit' and refers to each distinct digit in a stream of 1s and 0s. The bit represents the fundamental unit of digital information and, as proved by Claude Shannon, can be used to measure the information content of any message.

There are basically two components to this. First is determining the length of a message in binary notation, because a longer message can contain more information. But this is clearly not enough, because the content of the message needs to be included, otherwise a long meaningless message could be deemed more informative than a short, meaningful message. So, the length of the message needs to be multiplied by the probability of that message being true or occurring: a meaningless message has no probability of occurring and so has no information content. The resultant formula developed by Shannon has helped guide the development of ever more efficient systems for designing and transmitting information, including the internet.

In computers, strings of bits are split into groups of eight, known as bytes, which represent the number of bits needed to encode a digital number or English letter. Bytes are now used as the unit of measurement for digital memory; so, a 200 gigabyte hard disk comprises 200 billion bytes or 1.6 trillion bits. Actually, that isn't entirely true because a gigabyte is actually 2^{30} bytes (or slightly over 1 billion), because we're still working in binary. It's close enough, though.

The microprocessor

Once Shannon showed that this was the case there was no turning back: the future was digital. As transistors shrunk and ever more could be fitted onto a single silicon chip (see Spotlight, below), scientists were able to design more complex circuits that could process these binary digits in ever more advanced ways. The microprocessor was born, with the most advanced versions currently containing around 10 billion transistors.

Furthermore, the results of this processing could be stored as binary digits by other transistors, producing memory. Binary digits could then travel between this memory store and the microprocessor. They could also be introduced from outside, in the form of a computer program able to control the connections between the different elements of the microprocessor. The modern computer was born.

Now computers are everywhere, and they are extraordinarily powerful: there is much more computing power in a modern smartphone than was used to put a man on the moon in 1969. They have transformed every aspect of our lives: from work to entertainment to communications to relationships. They have allowed us to generate and process unprecedented amounts of information, providing unparalleled insights into the workings of the universe (see Chapter 25) and leading to the rise of artificial intelligence (see Chapter 27) – in addition to providing access to all the funny cat videos we could ever want to view.

Spotlight: Law abiding

In 1965, Gordon Moore, later to become co-founder of Intel, the famous computer chip company, came up with a law, which quickly became known as Moore's law. It's not really a law, though, more of a self-fulfilling prophecy. Moore's law states that the number of transistors that can be fitted on to a computer chip will double every two years.

Because more transistors mean faster and more powerful computer chips, it is the reason why we now have smartphones with far more computing power than was used to put a person on the Moon in 1969. For despite initially being little more than guesswork, Moore's law has basically held true for over 50 years, leading computers to become millions of times more powerful. And the reason it has held true is because the computer industry has treated Moore's law as a challenge, working hard to ensure the number of transistors that can be fitted on to a computer chip does indeed double every two years.

Actually, Moore's first iteration of his law stated that the number of transistors doubled every year, but that proved to be a bit too ambitious and so he changed it to every two years in the 1970s. But even this version is proving increasingly difficult to stick to. Not only is this due to the difficulty of fabricating transistors at ever smaller sizes, with state-of-the-art chips in 2019 containing transistors that are just 10 nanometres (billionths of a metre) wide. But also because transistors become electrically leaky at such small scales, preventing them being turned properly on or off (see Chapter 27).

Computer scientists are investigating various ways to stay within the law. This includes taking advantage of new designs, such as three-dimensional stacked chips, and new materials, such as graphene and other so-called two-dimensional materials (see Chapter 19). They are even developing entirely new computing technologies, such as spintronics, in which 0s and 1s are encoded in a property of electrons known as 'spin' rather than in their charge (see Chapter 25).

Key ideas

- Because the inputs and outputs in Boolean algebra can only adopt two values, true or false, they can be represented by electrical switches.
- A transistor is a tiny electrical switch made from a semiconducting material based on silicon or germanium.
- Multiple transistors can be joined together to perform specific logical functions, producing what are known as logic gates, such as AND, NOT and OR gates.
- The most efficient way to encode, store and transmit any information is as binary digits.
- Moore's law states that the number of transistors that can be fitted on to a computer chip will double every two years.

Dig deeper

Gleick, James, *Information: A history, a theory, a flood* (London: Fourth Estate, 2011).

Soni, Jimmy and Goodman, Rob, *A Mind at Play: How Claude Shannon invented the information age* (New York: Simon & Schuster, 2017).

19

Material advance

Sticky tape hasn't taken part in many major scientific discoveries, but it has been responsible for one. On a Friday evening in 2004, two scientists at the University of Manchester in the UK, Andre Geim and Kostya Novoselov, were playing around with a lump of graphite, using sticky tape to pull layers of graphite from the lump.

Graphite is an allotrope of carbon, meaning that it is a material made entirely of carbon atoms, as is diamond. But because the carbon atoms are arranged in different configurations in graphite and diamond, they are very different materials with very different properties.

In diamond, each carbon atom is bonded to four others to form a tetrahedron shape, making diamond the toughest known natural material and also attractively sparkly. Graphite, on the other hand, consists of lots of layers of carbon atoms, in which each carbon atom is bonded to three others to form a flat, hexagonal lattice like chicken wire. Because it consists of lots of layers, graphite is soft enough to be used as a lubricant, as well as being dark rather than sparkly and able to conduct electricity.

Graphite and graphene

Using their sticky tape, Geim and Novoselov eventually succeeded in extracting a single layer of carbon, just one atom thick, from the lump of graphite, creating another allotrope of carbon termed 'graphene'. This novel allotrope had been predicted, but no one had managed to isolate it before. The real excitement came, however, when Geim and Novoselov tested the properties of graphene, because this revealed it to be transparent, flexible and very strong; in fact, it's the strongest material ever measured. It can also conduct electricity better than copper.

But graphene turned out to be just the tip of the iceberg, because its discovery spurred scientists, including Geim and Novoselov, to look for other materials that consist of a single atomic layer, and they ended up finding quite a few. Because these materials are essentially all surface, having no effective depth, they are termed 'two-dimensional' (2D) materials.

Two-dimensional materials

2D materials include versions of graphene made with different elements, all of which have the characteristic '-ene' suffix, such as silicene, based on silicon, and phosphorene, based on phosphorus. They also include molecular versions, which consist of a single layer of molecules rather than a layer of atoms. Examples include boron nitride, also known as white graphene, which shares the hexagonal configuration of graphene but comprises alternating boron and nitrogen atoms, and molybdenum disulphide, in which a layer of molybdenum atoms is sandwiched between two layers of sulphur atoms.

Some of these 2D materials, such as molybdenum disulphide, can be extracted from a layered bulk material. Others are produced from the bottom up by chemical processes such as chemical vapour deposition, in which gases containing the compounds that make up the material are passed over a surface. Under the right conditions, the gases react at the surface to produce the desired 2D material. Some, such as graphene, can be produced by both methods.

Between them, these 2D materials have various different properties. Whereas graphene is a very effective conductor of electricity, hexagonal boron nitride is an insulator, meaning it's very poor at conducting electricity. Molybdenum disulphide is intermediate between the two, conducting electricity less well than a conductor such as graphene but better than an insulator, making it a semiconductor.

HETEROSTRUCTURES

Now scientists are investigating what happens when they stack different 2D materials together to form heterostructures. If you stack lots of graphene layers together, then you get bulk graphite, and if you stack lots of molybdenum disulphide layers together, then you get bulk molybdenum disulphide. However, if you stack graphene layers and molybdenum disulphide layers together, then you get a material that doesn't exist in nature and likely possesses some very interesting properties.

For that is the whole point of all this research effort: to develop materials with novel properties that can do things existing materials can't, allowing us both to improve our current technologies and come up with entirely new ones. Graphene and other 2D materials, together with the heterostructures made from them, are already being investigated for use in next-generation computer circuits (see Chapter 18), solar cells, batteries and biological sensors.

Advanced materials

Such materials are just the tip of an even bigger iceberg, because improvements in synthesis processes, analytical techniques and computer modelling are driving the development of many new materials. This used to be a trial-and-error process, with scientists simply reacting different compounds together to see what materials were produced, guided by little more than hunches.

Now that scientists can probe the atomic structure of materials with the latest analytical techniques and produce detailed computer models of them, they are beginning to understand much more about these materials and their properties. And this understanding is allowing them to focus their efforts, leading to the creation of whole new classes of advanced materials.

HIGH-ENTROPY ALLOYS AND PEROVSKITES

These include high-entropy alloys, which contain four or more different metallic elements in similar proportions – unlike conventional metallic alloys such as steel, which comprise one main element with one or two other elements as minor components. Steel, for example, is mainly iron with a tiny bit of carbon. High-entropy alloys can be stronger and tougher than conventional alloys, able to operate at higher temperatures and withstand corrosion better.

There is a crystalline material known as a halide perovskite that is showing great potential as a replacement for silicon in solar cells. Halide perovskite is the name for a group of materials with the same basic chemical structure, comprising an organic cation, an inorganic cation and a halide anion such as iodide, bromide or chloride in specific proportions. This means that by utilizing different anions and cations scientists can produce a wide range of halide perovskites with different properties.

Like silicon, several halide perovskite materials are able to convert sunlight into electricity. Unlike silicon, however, halide perovskites are flexible and can theoretically be manufactured using the same roll-to-roll process that produces newspapers, potentially making them cheaper and more versatile than silicon solar cells.

TOPOLOGICAL INSULATORS

Then there are highly unusual materials called topological insulators. Conventional materials can be conductors, insulators or semiconductors, depending on how well they conduct electricity. But a single material is always just one of these.

That's not the case for topological insulators such as bismuth telluride, which are insulators in their interiors but conductors at their surfaces. Indeed, topological insulators appear to be such good conductors at their surfaces that scientists are investigating whether they could form the basis for high-temperature superconductors. Unlike normal conductors, superconductors are able to conduct electricity without any loss of current due to heat, potentially leading to much more efficient power lines and electric motors. All known superconductors, however, work only at very low temperatures, hundreds of degrees below zero, hence the interest in developing versions that can work at higher temperatures.

Topological insulators could also potentially find use in 'spintronic' computing (see Chapter 18) and quantum computing (see Chapter 27). This is because topological insulators are examples of quantum materials, gaining their unusual properties from the quantum behaviour of subatomic particles, which in this case are electrons.

Graphene is another quantum material, because its atomic scale thickness means it is influenced by quantum effects (see Chapter 25), which give graphene its impressive properties. The same is true for other nanomaterials.

Nanomaterials

As their name implies, nanomaterials are materials with at least one dimension at the nanoscale, meaning a size of 1–100 nm; this is the scale of molecules and proteins. Scientists have developed ways to produce a whole range of nanomaterials with defined shapes and structures. And these nanomaterials often possess very different properties to the same materials at larger scales.

GOLD

Take gold. Now, the one thing that everyone knows about gold is that it's gold coloured and the one thing that every chemist knows about gold is that it's inert, meaning that it doesn't readily take part in chemical reactions. But all that changes if you transform a lump of gold into countless nanoparticles.

Then, gold will shine a range of different colours, with the precise colour (or frequency) of the emitted light depending on the size and shape of the nanoparticle. For example, rod-shaped gold nanoparticles just 20 nm wide and 60 nm long emit bright red light. This is known as fluorescence and normal gold doesn't do it.

Gold nanoparticles are also catalytic, meaning they can speed up certain chemical reactions without actually taking part in the reaction. A chemical reaction usually requires some form of energy, often provided in the form of heat, to take place, but catalysts reduce the amount of energy required, essentially by giving the reaction a helping hand. Gold nanoparticles are able to catalyse oxidation reactions, which in their simplest form involve adding oxygen atoms to a molecule.

This is an amazing transformation for a formerly dull, inert substance; like a shy person at a party who becomes the life and soul after a few glasses of wine. It is a result both of gold nanoparticles having a much greater surface area than bulk gold (see Spotlight), offering more surface for interactions, and of the fact that quantum effects become much more important at tiny scales. These effects are simply too weak to make much of an impact at large scales, but at the nanometre scale they allow gold to fluoresce and catalyse reactions.

Spotlight: Surface area explained

That smaller particles have a larger relative surface area can be shown with a simple box. Say you have a square box with sides 10 cm long; this box will have a volume of 1,000 cm^3 (10 × 10 × 10) and a total surface area of 600 cm^2 (10 × 10 × 6 sides).

Now say you replace this box with two boxes exactly half the size, so that each box has a volume of 500 cm^3. Each box will have sides 7.94 cm long (7.94 × 7.94 × 7.94 = 500) and a surface area of 378 cm^2 (7.94 × 7.94 × 6 sides), producing a combined surface area of 756 cm^2 for both boxes.

So simply replacing a single box with two boxes half the size increases the total surface area by more than 25 per cent. Also, the ratio of surface area to volume is larger for each of the two small boxes (0.756:1) than for the single large box (0.6:1)

As you split a set amount of material into smaller and smaller particles, the combined surface area of the particles gets larger and larger, as does the ratio of surface area to volume for each individual particle.

BUCKYBALLS AND NANOTUBES

Graphene is only one of several carbon-based nanomaterials. There are buckyballs, which are spherical cages made of 60 carbon atoms, and carbon nanotubes, which are tubes of carbon atoms, as though a layer of graphene had been rolled up. Indeed, another method for producing graphene is to effectively cut carbon nanotubes up the middle and lay them flat. Unfortunately, turning the graphene back into nanotubes requires more than just a bit of sticky tape.

Spotlight: Printing enters a new dimension

Scientists are not just developing new materials, but also new ways to process existing materials. Conventional production processes tend to involve transforming a big load of some metal or plastic material into lots of smaller products, whether by cutting or moulding. Such processes are wasteful, tending to generate a lot of cast-off material, and also inflexible, because they're only cost effective if they produce lots of exactly the same product, known as mass manufacturing.

Three-dimensional (3D) printing is very different. Also known as additive manufacturing, it involves building up a product from the bottom up, layer by layer, according to a computer-based design. There are various ways to do this. The layers can be deposited as molten plastic, which solidifies as it cools, or as a metal powder that is fused together with a laser beam. Or the product can emerge from a vat containing a light-responsive liquid polymer, which solidifies when illuminated with a laser beam shone in a defined pattern.

Whatever the method, 3D printing is far less wasteful, because it doesn't generate cast-offs and any leftover material can simply be reused. It's also much more flexible, because a product can easily be tweaked or customized by simply changing the computer-based design. And because it prints objects layer by layer, 3D printing can realize complex designs that simply can't be produced by conventional production processes.

Nevertheless, the age of 'mass customization' promised by 3D printing has not quite usurped the age of mass manufacturing, mainly because 3D printing remains a slower and more expensive way to manufacture products en masse. But it is beginning to find its niches. For example, its flexibility is allowing 3D printing to produce shoes made for the shape of individual feet and its ability to produce complex shapes is allowing it to print lighter aircraft parts. The falling cost of commercial 3D printers is also transforming many other fields, including science, where it offers scientists a quick and easy way to produce instruments and devices.

Key ideas

- Graphite and diamond are allotropes of carbon, meaning they are both made entirely of carbon atoms but in different configurations.
- Graphene is a one-atom-thick layer of carbon atoms that is transparent, flexible and very strong, and can conduct electricity better than copper.
- Improvements in synthesis processes, analytical techniques and computer modelling are leading to the creation of whole new classes of advanced materials.
- Quantum materials gain unusual properties from quantum effects or from the quantum behaviour of subatomic particles.
- Nanomaterials are materials with at least one dimension at the nanoscale, meaning a size of 1–100 nm.

Dig deeper

Clegg, Brian, *The Graphene Revolution: The weird science of the ultra-thin* (London: Icon Books, 2018).

Miodownik, Mark, *Stuff Matters: The strange stories of the marvellous materials that shape our man-made world* (London: Penguin Books, 2014).

20

It's alive

Humankind has been taking advantage of biology to create desirable products for thousands of years – using yeast to produce bread and beer, for example – but we've always been limited by what the organisms were naturally able to do. Yeast can convert sugar into alcohol and produce the carbon dioxide that makes bread rise, but it can't convert sugar into petrol.

With modern genetic techniques, scientists are starting to design strains of yeast that can do just that. Or that can produce other useful chemical products, such as drugs or plastics, or that can clean up pollution such as oil spills.

We are now entering the stage where we can program and customize life in much the same way that we can program and customize computers and cars. We are able to tinker with organisms, modify them and bend them to our will. We can make them do whatever we want, for both good and ill.

The path to modern genetics

TRANSFERRING GENES

The work that led us to this point goes back over 40 years to when scientists first transferred single genes between species. As we saw in Chapter 6, genes provide the instructions for making proteins, which are both the building blocks and catalysts of life. An organism's full collection of genes, known as its genome, provides the complete instruction manual for building and operating that organism.

Although most individual organisms have their own unique genome, each genome is made up of exactly the same four DNA bases, simply arranged in a different sequence. That means that a gene from one organism should work just as well in another organism, as the underlying programming language is the same. If that gene codes for a protein that provides the original organism with a useful ability, then transferring that gene to another organism should also transfer the ability.

USING PLASMID VECTORS

There are two main ways to transfer genes between organisms: plasmid vectors and viral vectors. Plasmids are circular strands of DNA found in many bacteria, existing and reproducing independently of the main bacterial chromosome. Bacteria naturally transfer plasmids to each other, and this is the main mechanism by which certain genetic abilities, such as resistance to antibiotics, pass quickly among bacteria.

In the early 1970s, scientists developed techniques for snipping out a gene from one organism's genome and incorporating it into a specially prepared plasmid, which can then be introduced into a bacterial, fungal or eukaryotic cell. Depending on the nature of the plasmid, it either remains independent of the host cell's chromosome or becomes incorporated within it. Either way, the plasmid can be designed such that the novel gene is expressed by the cell's protein-constructing machinery, providing the cell with the novel ability.

The tricky part is getting the plasmid into the cell, but there are a number of ways this can be done. Bacteria are fairly easy, happily picking up any plasmids added to their growth medium, but fungi and eukaryotic cells tend to be a bit pickier. Getting plasmids into these cells requires either opening up holes in their cell wall or outer membrane, which can be done by applying a short electric shock, or inserting the plasmids inside vehicles that the cells will happily take up, such as fatty bubbles known as liposomes. Alternatively, the plasmid can be attached to microscopic metal particles and literally fired into the cell.

USING VIRAL VECTORS

This whole process can be circumvented by using viral vectors. In this case, the novel gene is added to viral DNA, as viruses are little more than a strand of DNA (or RNA) surrounded by a protein shell. But viruses are very good at getting their DNA inside cells, by essentially injecting them through the cell membrane or wall. Once inside, the viral DNA either incorporates itself into the cell's genome or tricks the cell's own protein-constructing machinery into converting its genes into proteins.

Because different viruses are specialized at infecting different kinds of cells, scientists utilize several viruses as the basis for viral vectors. These viruses have been modified to enhance their ability as a gene-delivery vector and to remove their ability to cause harm, although this is not always fool-proof.

GENETICALLY MODIFIED (GM) ORGANISMS

Using plasmid and viral vectors, scientists have over the past 40-plus years created a whole menagerie of organisms with foreign genes, known as genetically modified (GM) organisms. Various GM microbes have been created to produce therapeutic proteins, including insulin, human growth hormones and human Factor VIII, which helps the blood to clot and is used to treat haemophilia. GM crop plants such as maize and soybeans, usually modified to be resistant to certain herbicides or to produce a natural insecticide, are now grown all over the world. GM animals are less widespread, but GM mice with human genes are now regularly used in laboratories to explore human diseases.

Cloning animals

Genetic technology has also continued to advance. Rather controversially, scientists are now able to clone animals, producing an exact genetic copy. Scientists are essentially creating an identical twin of an organism that has already been born.

To do this, they take an egg, meaning a female gamete (see Chapter 8), and remove its nucleus, replacing it with the nucleus of a cell from the animal being cloned, say a skin cell. Then they give the egg a short electric shock to get it to start dividing. After it has divided for four or five days, becoming a very young embryo known as a blastocyst, it is implanted into the womb of a female member of the species. Sometime later, a clone is born.

The first cloned mammal was the famous Dolly the sheep, born in 1996 (see Spotlight), but since then various other mammals have been cloned, including pigs, horses and dogs. Cloning itself has only limited practical benefits, but the same technique could also be used to produce personalized stem cells, which have great medical potential.

Spotlight: Dolly the sheep, 1996–2003

Produced by Ian Wilmot and his team at the Roslin Institute near Edinburgh in Scotland, Dolly was the first successfully cloned mammal. Using the technique described in the main chapter, Dolly was cloned from a mammary gland cell taken from a six-year-old sheep and so was named after the famously busty country singer Dolly Parton.

Dolly demonstrates the difficulties and pitfalls of cloning a mammal. For a start, she was the only lamb to make it to adulthood from 277 cloning attempts. She also died young for a sheep, which usually have a life expectancy of around 12 years, suffering from lung disease and acute arthritis.

It's still not clear, however, whether these medical conditions were exacerbated by her being a clone. It has been suggested that producing a clone from an adult cell could lead to a shorter life, simply because adult cells have already undergone some ageing.

Stem cells

EMBRYONIC STEM CELLS (ESCS)

Stem cells are body cells that can reproduce indefinitely. Our bodies contain many different types of stem cell, which create new cells to replace those that have died off, but most stem cells can only produce a limited range of cell types. The gold standard of stem cells is embryonic stem cells (ESCs), which, as their name suggest, are only found in young embryos.

ESCs are cellular blank slates, able to produce any one of the 220 different types of cell found in the human body. They do this by repeatedly dividing, producing both more stem cells and cells that are increasingly specialized, eventually becoming a specific cell type.

Because they have this ability, ESCs are described as pluripotent; less flexible stem cells that produce a smaller range of cells are described as multipotent or unipotent.

In so-called therapeutic cloning, scientists would use cloning technology to create blastocysts that are genetically identical to a patient suffering from some kind of degenerative disease. ESCs are then extracted from these blastocysts and used to produce specific cell types or even whole organs in the laboratory, which are then transplanted into the patient. Because these cells and organs will be genetically identical to the patient's own cells, their immune system shouldn't attack them. If the ESCs were turned into neurons, for instance, then they could be used to treat Alzheimer's disease, which is caused by the loss of large numbers of neurons (see Chapter 10).

In 2013, after many years of work in this area and an earlier false claim of success (see Chapter 21), a group of scientists succeeded in producing the first cloned human ESCs. By this time, though, interest had mainly shifted to a much less controversial stem cell technology.

INDUCED PLURIPOTENT STEM (IPS) CELLS

Therapeutic cloning is controversial because it involves destroying human blastocysts. Even though blastocysts comprise collections of a just few hundred cells, many people find the idea of creating embryos just to destroy them morally questionable. In 2006, however, scientists discovered that certain adult cells, including skin and fat cells, could be converted into cells with many of the same properties as ESCs by simply introducing three or four specific genes into them.

Known as induced pluripotent stem (iPS) cells, they offer a far less contentious way to produce cells that are genetically identical to a specific patient and shouldn't be rejected by their immune system. Scientists have already shown that iPS cells can be differentiated into various cell types by exposing them to specific combinations of proteins and other molecules, just like ESCs. These include liver cells, stomach cells and neurons, although scientists have yet to try transplanting these cells into patients.

They have, however, succeeded in inducing iPS cells to grow into basic versions of organs, known as organoids, including brain and intestine organoids. These can reveal information about the development of the actual organ and its response to disease. Other scientists have found a shortcut to using iPS cells to produce an organ for potential transplant into patients. Rather than grow the organ from scratch, they take an embryo from an animal such as a pig, and switch off the genes that promote the development of an organ of interest, such as the heart, using a novel gene editing technique (see Spotlight). Next, they inject iPS cells derived from a patient who needs a new heart into the embryo, which is then implanted into the womb of a sow.

The idea is that, as the embryo develops, it is forced to utilize the genes in the iPS cells to produce a heart, as its own heart genes have been switched off. This means the resulting pig should have a heart made of human cells that are genetically identical to the patient, making it ideal for transplant. This work is still at a very early stage, with scientists currently finding that pig organs produced in this way contain only a tiny proportion of human cells.

Spotlight: Gene editing gets CRISPR

Scientists have long been able to insert new genes into cells and switch off existing genes, but a new gene editing technique with the catchy name of CRISPR-Cas9 offers the ability to do this faster and more precisely than ever before.

CRISPR stands for 'clustered, regularly interspaced short palindromic repeats', which is a region of DNA found in the genomes of many bacteria, while Cas9 is the name of an enzyme that can slice through DNA strands. In bacteria, the CRISPR region codes for RNA strands able to bind with the DNA of viruses prone to infecting bacteria. This binding creates a complex targeted by Cas9, which slices through the viral DNA, deactivating it.

Bacteria use CRISPR-Cas9 to protect themselves against viral infection, but scientists realized they could adapt the process for gene editing. The idea is to insert into a cell an RNA strand designed to bind to a specific gene in the cell's genome, together with Cas9. This binding again forms a complex that is targeted by Cas9, which slices through the gene, deactivating it. Scientists also found that if they inserted a novel gene into the cell at the same time, it becomes incorporated into the genome at the site where the old gene has been deactivated. Previously, scientists had struggled to insert genes into specific places in a genome.

CRISPR-Cas9 is generating a lot of excitement because of its potential for treating a wide range of diseases. In particular, diseases such as sickle cell anaemia that are caused by a single faulty gene, which could be replaced by a working version with CRISPR-Cas9. This editing would be most effective in early embryos, so the genetic fix is carried through to all the growing body's cells. Nevertheless, until questions regarding the safety and ethics of the technique are resolved, scientists have called for a moratorium on this work in humans. In 2018, however, a Chinese scientist announced, to widespread condemnation, that he had used CRISPR-Cas9 to edit the genomes of two embryos that were subsequently born as twin girls.

Synthetic biology

Another recent genetic advance is that scientists can now insert suites of foreign genes into organisms rather than just single genes, allowing them to transfer more complex traits and abilities. They can also enhance the activity of existing genes, making them produce more protein, or reduce their activity, as well as easily and accurately switch existing genes off and add new genes using the new gene editing technique. This 'extreme' genetic engineering is known as synthetic biology and allows scientists to provide organisms with far more advanced abilities. For instance, scientists have already developed GM organisms that can convert sugars into certain industrial chemicals and fuel-like compounds. This work is being boosted by the increasing number of organisms that have had their genomes sequenced (see Chapter 7), because it provides scientists with a bigger pool of known genes.

Scientists can also chemically synthesize all four DNA bases and join them together into strands, which means specific DNA sequences can now be produced to order. In 2010, this allowed scientists to create the world's first artificial genome, made from synthetic DNA, which they transplanted into a bacterial cell, replacing the bacterium's own genome.

Although this feat was promoted as the creation of the first synthetic life form, the artificial genome was an almost exact copy of a natural bacterium genome. Since then, however, the same group of scientists has been removing genes from this synthetic genome to find the smallest version that can still keep a bacterium alive.

The moral dilemma

The aim now is to produce genomes that aren't simply replicas of natural genomes but are entirely novel constructs that provide the host organism with entirely novel abilities, such as being able to convert sugar into petrol. This is being assisted by the wholesale automation of genetic technology, such that the latest laboratories can create and test 15,000 different genetic designs a day.

And just as with stem cells, this new ability to program and customize life is throwing up a whole host of moral and ethical questions. We really have entered a brave new world.

Key ideas

- A gene from one organism can work just as well in another organism, as the underlying programming language is the same.
- Over the past 40-plus years, scientists have created a whole menagerie of organisms with foreign genes, known as genetically modified organisms.
- In therapeutic cloning, scientists apply cloning technology to embryonic stem cells (ESCs) to create biological tissue that is genetically identical to a patient.
- Scientists have discovered that certain adult cells can be converted into cells with many of the same properties as ESCs, called induced pluripotent stem cells.
- With synthetic biology, scientists can now transfer and edit whole suites of genes, allowing them to introduce more complex traits and abilities.

Dig deeper

Metzl, Jamie, *Hacking Darwin: Genetic engineering and the future of humanity* (Chicago: Sourcebooks, 2019).

Mukherjee, Siddhartha, *The Gene: An intimate history* (London: Vintage, 2016).

Part Five

When science goes bad

21

Fraud, fakery and fantasy

In October 2009, the South Korean stem cell scientist Woo Suk Hwang was convicted of embezzling research funds and illegally buying human eggs. He was given a two-year suspended prison sentence.

His scientific career was already in tatters, following the retraction of two research papers that had appeared in the journal *Science* in 2004 and 2005. These papers detailed Hwang's ground-breaking stem cell work. In the first paper, he claimed to have derived embryonic stem cells from a cloned human blastocyst, the holy grail of therapeutic cloning (see Chapter 20). In the second paper, he claimed to have extended this work by producing embryonic stem cells from tissue contributed by patients suffering from disorders such as diabetes and spinal cord injury, potentially offering a way to cure these and other currently intractable medical conditions.

For this work, Hwang was feted around the world, becoming a national hero in South Korea. The problem was a lot of it was fake; together with certain members of his research team at Seoul National University, Hwang had fabricated much of the data. The stem cells either did not exist or were not derived from cloned blastocysts. Hwang also paid for many of the human eggs used in the work, some of which came from his own female researchers; a major breach of medical ethics.

More cases of scientific fraud

Other scientists convicted of scientific fraud and misconduct have gone straight to jail. In 2006, Eric Poehlman, a former menopause and obesity researcher at the University of Vermont in Burlington, became the first US scientist to go to prison for scientific misconduct where there were no patient deaths. This followed his admission that he had falsified data in 15 grant applications and numerous research papers, including simply making up patients.

In 2015, Dong-Pyou Han, a former biomedical scientist at Iowa State University in Ames, US, was sentenced to 57 months in prison and fined $7.2 million for fabricating and falsifying data in trials of a vaccine against the human immunodeficiency virus (HIV), which causes AIDS. This included spiking samples of rabbit blood with antibodies to human HIV, to give the impression that the vaccine had caused the animals to develop immunity to the virus.

SCIENTIFIC FRAUD IN OTHER AREAS OF SCIENCE

Scientific fraud is not just confined to biomedicine, although it has been suggested that it is more prevalent in this area of science. This may be because the stakes, in terms of financial gain and prestige, tend to be higher here than in less immediately practical scientific fields, encouraging fraudulent claims. Or it may simply be the case that the strong emphasis on openness in biomedicine makes it more likely that cheats will be exposed.

But fraud has been uncovered in almost all areas of science, including chemistry, physics, environmental science, psychology and archaeology (see Spotlight, below). For example, another major fraud occurred in organic electronics, the field of developing electronic and computing devices from carbon-based material such as plastics rather than silicon.

In the late 1990s and early 2000s, Jan Hendrik Schön, a young physicist at the prestigious Bell Laboratories in New Jersey, US, published a string of papers reporting major advances in organic electronics, including the first organic electrical laser. He also reported producing a superconductor from the carbon nanomaterial known as buckyballs (see Chapter 19). Most superconductors only work at very low temperatures, but Schön claimed that his buckyball superconductor worked at much higher temperatures. In 2000 alone, Schön published eight papers in *Science* and *Nature*, the two most prestigious scientific journals.

For these breakthroughs, Schön was garlanded with awards and universally acknowledged to be on course for a Nobel Prize. Unfortunately, it was all a lie; Schön hadn't achieved any of the breakthroughs that he claimed. Suspicions were raised when other research groups couldn't replicate his work, but what finally did for him was that he used exactly the same graph to illustrate different findings in different papers. When this all came out in 2002, Schön was immediately fired by Bell Laboratories.

Spotlight: The missing link that wasn't

Perhaps the most famous scientific fraud of all time, so immense that it is described as a hoax rather than a fraud, is Piltdown Man. In 1912, a British solicitor and amateur archaeologist called Charles Dawson announced the discovery of fragments of an ancient skull that appeared to be that of a missing link between apes and humans. It was termed Piltdown Man, after the small Sussex village where the skull was supposedly unearthed.

Almost immediately, some scientists questioned the skull's authenticity, but for the next 40 years it was generally accepted as the skull of an early human. Subsequent findings of real early human remains, however, indicated that Piltdown Man was, at the very least, an anomaly in the evolution of man.

Then in 1953 the latest analytical techniques finally proved that the skull was a forgery. It actually comprised a jawless human skull from the medieval era, attached to which was an orangutan jaw with the teeth filed down. The perpetrator of the hoax is still unknown, although much of the evidence points to the supposed discoverer, Charles Dawson.

Data manipulation

Although such major cases of scientific fraud are fairly rare, less extreme examples of data manipulation and fabrication are probably more common. Reliable figures are hard to come by, as scientists are understandably not keen to admit their misdemeanours, but a number of anonymous surveys of such activities have been carried out over the years.

In 2009, a researcher at the University of Edinburgh in Scotland reviewed many of these surveys, finding that overall just 2 per cent of the scientists questioned admitted to fabricating, falsifying or altering data at least once.

More worryingly, however, 34 per cent of scientists admitted to other questionable research practices, including 'failing to present data that contradict one's own previous research' and 'dropping observations or data points from analyses based on a gut feeling that they were inaccurate'. Furthermore, when asked to report on the behaviour of their colleagues, 14 per cent said they knew someone who had fabricated, falsified or altered data, and over 70 per cent knew someone who had committed other questionable research practices.

Another way to gauge the extent of scientific fraud is from the number of published research papers that are subsequently retracted because their findings don't hold up to scrutiny. A study in 2018 reported that, on average, just four papers in every 10,000 are retracted. But that still meant 946 papers were retracted in 2014, of which just over 40 per cent were retracted due to fraud (the others were retracted due to errors and other problems). Furthermore, a study of 20,000 papers in 2016 revealed that 2 per cent contained 'problematic' scientific images that may have been deliberately manipulated.

So what exactly is going on? Well, part of the problem is that the world is a messy and noisy place. Scientists are often looking for a small, specific effect in the midst of numerous other variables and unavoidable fluctuations, termed noise. Coming up with definitive evidence for that effect can be difficult and often requires complex statistical analysis to separate out all the other variables and noise.

Sometimes scientists simply overstep the mark, removing or changing inconvenient data to make the world behave as they think it should and to give them the results they want. It's like a golfer surreptitiously throwing his golf ball out of the rough and on to the fairway because he knows that's where it should really have landed.

Also, scientists may like to think they're impartial searchers after truth, but in reality they are often motivated by more worldly considerations, like money, prestige, fame, awards and career advancement – as are we all. For scientists, these rewards tend to come from making important scientific findings and advances, and then publishing research papers about them in prestigious scientific journals like *Science* and *Nature*. Hence, the incentive to come up with these findings, even if the data don't quite warrant them.

Still, it's a long way from a small bit of data manipulation, even though this is still highly undesirable, to major fraud. But most scientists don't set out to commit fraud, rather they believe that their work is genuine and just can't accept that they have not quite got the results they wanted.

The discovery of 'cold fusion'

Most scientific fraud starts out merely as wishful thinking. For example, Hwang was apparently convinced that the first stem cells he produced were derived from cloned blastocycts, even though it subsequently turned out that they weren't. In its most extreme form, however, even wishful thinking can get scientists into bother.

In 1989, the world's energy troubles seemed to be over, after two chemists at the University of Utah in the US, Stanley Pons and Martin Fleischmann, announced the discovery of 'cold fusion'. As we saw in Chapter 2, nuclear fusion takes place in the centre of stars, where atomic nuclei fuse together to form larger nuclei, releasing huge amounts of energy in the process.

Scientists have long dreamt of replicating this process on Earth, by fusing deuterium isotopes together to form helium. Deuterium is an isotope of hydrogen, with a nucleus comprising one proton and one neutron (see Chapter 2). It is naturally found in the oceans, where it forms what is known as heavy water, due to its extra neutron. The problem is that a sustainable fusion reaction requires immense temperatures and pressures, which are naturally generated in the centre of stars but are harder to sustain for long periods on Earth.

In 1989, however, Pons and Fleischmann announced they had produced fusion in a tank of heavy water at room temperatures, making headlines around the world. Their method involved simply passing electricity between two palladium electrodes immersed in heavy water, in a version of a well-known technique called electrolysis. But the chemists claimed that their version produced neutrons and much more energy than could be explained by normal chemical reactions. They reasoned that deuterium nuclei absorbed by the negative electrode were being squeezed together so tightly that they fused, generating the energy and neutrons.

BYPASSING THE SCIENTIFIC JOURNALS

The problem was that not only did this breakthrough contradict the known laws of physics, but Pons and Fleischmann didn't announce it in *Science* or *Nature* but at a press conference. This was a major breach of scientific etiquette, because new scientific research is usually first revealed to the world as a paper in a scientific journal.

This allows the research to undergo peer review, during which several experts review the paper to ensure it doesn't contain any major mistakes or omissions, and it also allows the research to be described in sufficient detail for other scientists to replicate it.

Pons and Fleishmann circumvented this process and they suffered for it. Although they didn't release the full details of their cold fusion method, they gave enough information at the press conference for other scientists to form a good idea of what they had done. These scientists then started to replicate their work, but were unable to replicate their findings.

Initially, Pons and Fleischmann argued that this was because of unspecified differences between their method and the replications. But as more and more scientists were unable to find any evidence of cold fusion, and as shortcomings with Pons and Fleischmann's own experiments became apparent, it became increasingly clear that the two chemists had simply been fooling themselves.

If Pons and Fleischmann had submitted their work to a journal, then these shortcomings would probably have been identified during peer review, saving the two chemists a lot of embarrassment. But peer review can only go so far: it can't always identify work that is deliberately fraudulent, because it's only meant to identify whether the experimental method is sound, not whether the scientists have invented the data.

Fortunately, this kind of fraud tends to be uncovered by the propensity of scientists to repeat each other's work, especially work that leads to major discoveries (see Spotlight). In many cases of scientific fraud, suspicions have first been raised by the inability of other research groups to replicate the findings.

In science, as in life in general, cheats don't tend to prosper.

Spotlight: Replicate and be damned

Scientists may be happy to try replicating ground-breaking discoveries, especially if they have suspicions about those discoveries, but they have tended to be much less keen on replicating more everyday studies. This is perfectly understandable: they get prestige and kudos for publishing novel research, not for simply trying to replicate other scientists' findings. Furthermore, scientific journals are keener on publishing novel research, and funding agencies are keener on paying for novel research. It's also not a problem, as long as the findings of the original studies are perfectly robust, but in a surprising number of cases they may not be.

This is being revealed by a few studies that have attempted to replicate the findings of a large number of research papers in a specific area of science. In 2012, scientists at Amgen, a US drug company, reported doing this for 53 landmark papers in cancer research, but were only able to replicate the original findings for six of them. In 2015, researchers tried to replicate the findings of 98 psychology papers, but could only do so for 39 per cent of them.

Rather than deliberate fraud, the natural variation and noisiness of the universe is probably to blame for this replication problem. As mentioned earlier in this chapter, the effects that scientists are looking for are often very small, revealed only by analysing their data with complex statistical techniques. A detected effect may simply be illusory, a random quirk of the data or the statistics. Or it may be real but so weak that sometimes it can be detected and sometimes it can't, even with ostensibly the same experiment.

Over time, as science progresses, anomalous findings will fall by the wayside, but they may still cause researchers to spend time, effort and money exploring futile dead-ends. Hence, the importance of the rather thankless task of replicating research, which is now starting to be taken more seriously. In 2016, an online journal was launched specifically for the purpose of reporting replications of previous studies.

Key ideas

- Fraud has been uncovered in almost all areas of science, including biomedicine, chemistry, physics, environmental science, psychology and archaeology.
- Although major cases of scientific fraud are fairly rare, less extreme examples of data manipulation and fabrication are probably more common.
- Scientists are often looking for a small, specific effect in the midst of numerous other variables and unavoidable fluctuations, which can be difficult.
- New scientific research is usually first revealed to the world as a paper in a scientific journal, which allows the research to undergo peer review.
- Scientific fraud tends to be uncovered by the propensity of scientists to repeat each other's work, especially work that leads to major discoveries.

Dig deeper

Park, Robert, *Voodoo Science: The road from foolishness to fraud* (Oxford: Oxford University Press, 2000).

Chevassus-au-Louis, Nicolas, *Fraud in the Lab: The high stakes of scientific research* (Cambridge, MA: Harvard University Press, 2019).

Shocks and scares

We live in a world of fear. Despite the fact that those of us living in the developed world are generally wealthier, healthier and longer lived than ever before, there seem to be no end of things for us to worry about. And many of these things have a technological bent: GM crops, man-made chemicals, overhead power lines, mobile phones, nuclear power and vaccines. Even breast implants have been known to give us the willies.

The irony is that we owe much of our increased wealth, health and lifespan to many of the technologies that we're now so concerned about. We don't appreciate what life used to be like just over 100 years ago: the widespread poverty, hunger and disease, the high proportion of women who died in childbirth and of children who died before their fifth birthday, and the far shorter lifespans.

We forget that technologies such as water treatment, vaccination, pesticides and fertilizers successfully banished all these real terrors. Instead, we take our wealth, health and long lives for granted and focus on the small, and in many cases non-existent, dangers posed by these life-saving technologies.

System 1 and System 2

But we just can't help thinking like this. Over the past few years, scientists have discovered that we are often simply unable to assess risks in a logical and rational way, and this is especially the case when those risks involve modern technology. The reason for this is that humans appear to possess two separate systems of thought, which scientists have rather unimaginatively labelled System 1 and System 2.

System 1 is essentially intuition or gut feeling. Its advantage is that it's quick and decisive, but it's not very deep; indeed, we are usually not aware of it at all. As such, when assessing a threat, System 1 does not dispassionately consider all the evidence before making a decision; that would take too long. Rather it uses some basic rules of thumb – what scientists call heuristics – to make a decision quickly.

In contrast, System 2 is all about dispassionately considering the evidence. It's logical and systematic, but it's also slow and takes a lot of mental effort. If we make a snap judgement about something based on a gut instinct, that's System 1. If we consider things for a long time, painstakingly weighing up the arguments for and against, that's System 2. As such, System 2 is responsible for all humankind's more impressive intellectual achievements, including science.

The problem is that the answers quickly reached by System 1 almost always colour the slower, deeper deliberations of System 2. That's if System 2 even gets involved, because in many cases it just goes with whatever System 1 has decided. Even if it does review System 1's decision, it only tends to alter or adapt it, rather than actually overrule System 1. So, although System 2 represents the logical, rational aspect of our thought processes, it's the fast and loose System 1 that tends to call the shots.

In order to make its speedy judgements, System 1 operates subconsciously, because conscious thought would slow it down, which means that we often don't know how we come to our decisions. That we are able to rationalize our snap judgements stems from the fact that System 2 can usually find plausible explanations for the decisions already provided by System 1. But it was the decisions that led to the explanations, rather than the other way around.

So, what are the subconscious rules of thumb, or heuristics, utilized by System 1? Well, the three main ones, determined from numerous behavioural experiments (see Spotlight), are termed the anchoring and adjustment heuristic, the representativeness heuristic and the availability heuristic. The anchoring and adjustment heuristic reflects our propensity for making an initial judgement using information we've just heard as a reference point. The representative heuristic reflects our propensity for equating typical events with likely events and the availability heuristic reflects our propensity for assuming that events that can be easily recalled are quite common.

Spotlight: Studying System 1

Because the heuristics and biases of System 1 operate unconsciously, scientists have had to design clever experiments to reveal their influence on our thought processes.

For example, in one study scientists asked groups of people whether the Indian political leader Mahatma Gandhi was older or younger than either 9 or 140 when he died. Now, these are obviously stupid questions, but when the scientists then asked the same people to guess how old Gandhi actually was when he died these questions influenced their answers. Those who were asked whether Gandhi was older or younger than 9 when he died guessed that he died at a younger age than those asked whether he was older or younger than 140. This is the anchoring and adjustment heuristic in action.

In a 1982 study, scientists asked groups of political experts meeting at a conference the likelihood that in the following year there would be either a complete suspension of diplomatic relations between the US and the Soviet Union or a Soviet invasion of Poland followed by a complete suspension of diplomatic relations. The experts rated the second scenario as more likely, even though logically the first scenario must be more likely, because the second scenario is simply a subset of the first scenario. This is the representative heuristic in action.

On top of this, System 1 also utilizes numerous 'biases' or tendencies in making its judgements. These include our tendency to be revolted by faeces and rotting material, and our tendency to seek out information that corresponds to our existing beliefs, known as the confirmation bias. Also, we tend to believe that risk and benefit are inversely related: so that risky things can't be beneficial and beneficial things can't be risky.

These heuristics and biases appear to be hard-wired in our brains, which means they are a product of evolution. And when modern man first appeared around 200,000 years ago, they were probably very valuable. At that point, if you were walking across the African savannah and saw a movement in the long grass ahead, you'd quickly need to decide whether it was being caused by a tasty antelope or a hungry lion. The heuristics and biases of System 1 could help you make that decision.

If you'd just heard someone mention a lion, if this looked like typical lion territory and if you could remember hearing about a recent lion attack, then it would probably be prudent to get the hell out of there as quickly as possible. If lion attacks weren't so high on the agenda, then it may be worth investigating further. With System 1 you wouldn't have to go through this laborious thought process, you'd just get a bad feeling and run away.

Because evolution is a slow process, System 1 is still using these heuristics and biases today, even though the world is very different. Still, System 1 can be useful when it applies its heuristics to an area where you have some expertise, allowing you to make judgements that are both quick and informed. Scientists have discovered that people working in highly stressful environments, such as fire-fighters and air-traffic controllers, often quickly make correct decisions without being able to explain how they came to those decisions. It's when System 1 applies its heuristics to areas where people are generally less knowledgeable, such as the risks of new technology, that it can reach some highly suspect conclusions, which System 2 often doesn't correct.

WHEN SYSTEM 1 MAKES US OVERLY WORRIED

System 1 is especially thrown by the fact that we now receive information from all over the world. We may logically know that seeing a report of an abducted child on the television news doesn't make it any more likely that our child will be abducted, but System 1 doesn't know that. It simply uses the availability heuristic to deduce that hearing about an abducted child means that abduction of children is quite common, no matter where that abduction took place. Hence, we become overly worried that someone will abduct our child while they play out in the street. Scientists have discovered that people consistently overestimate the likelihood of being killed by the kind of things reported in newspapers and on the television news, such as murders, floods and fires.

The same kind of dodgy thinking lies behind many of the technological health scares of the past few years. Despite the fact that almost all of these health scares turned out to have no basis in reality, they all possess qualities that cause System 1 to sound the alarm bells.

GM crops

GM crops have been widely grown and consumed for over 20 years without any detrimental effects on the environment or people's health; in 2017, 190 million hectares of GM crops were planted in 24 countries around the world. But publicity over a few highly contentious studies at the end of the 1990s that found ill effects in animals that ate certain GM produce, together with the fact that the first GM crops only benefited farmers rather than consumers, was sufficient to turn Europeans against GM crops. All these years later, GM crops are still not grown in most of Europe.

Stoked by environmental groups, people are also concerned about the concentrations of man-made chemicals in the environment, whether deliberately released chemicals such as pesticides or industrial chemicals that have simply escaped into the environment. Tests on people living in the developed world have shown that their blood contains numerous man-made chemicals, some of which are toxic or potentially cancer-causing.

Now this sounds alarming, but it's not really much cause for concern. Although all our bodies are contaminated with man-made chemicals, for most people they're at such low concentrations that they won't cause us any harm. Furthermore, many of the chemical compounds produced by nature, including some that end up in our food, are toxic or potentially cancer-causing at high enough concentrations, but no one worries too much about these natural chemicals.

The American Cancer Society has estimated that only around 2 per cent of all cancers are the result of exposure to man-made or naturally occurring environmental pollutants. In contrast, lifestyle factors such as smoking, drinking, diet, obesity and exercise have a much more important influence on the chances of developing cancer. But people will worry about the effects of miniscule traces of environmental pollutants in their body, while happily smoking, drinking and eating cream cakes.

Spotlight: But what about the environment?

Although the risks to our health from man-made chemicals tend to be fairly minimal, the risks to the environment can be much more serious.

For example, nitrogen-rich fertilizer applied to fields in the US mid-west regularly wash into nearby rivers, eventually ending up in the Gulf of Mexico. Every spring and summer, this influx of fertilizer stimulates the growth of huge algal blooms that suck all the oxygen out of the water. The end result is a massive dead zone, covering 1.5 million hectares of the Gulf, where nothing can live.

The widespread application of pesticides, meanwhile, has been blamed for dramatic reductions in the numbers of honeybees and other insects in many regions of the world.

Safety fears have also been raised about mobile phones, where the concern is that electromagnetic fields generated by the phones may cause cancer. But, unlike with GM crops, these fears have not stopped people from using their phones.

All these rather irrational responses to technology can be traced back to the heuristics and biases employed by System 1. Any news reports raising doubts about the safety of a new technology, even if these doubts are later shown to be unfounded, are enough to set off the heuristics utilized by System 1, immediately giving people a bad feeling about the technology.

Furthermore, because of the risk–benefit bias, people are more willing to accept that technologies which don't seem to have a direct benefit for them, such as GM crops, are risky. Whereas they're less willing to accept that directly beneficial technologies such as mobile phones can present a danger.

The MMR scare

Sometimes, these irrational responses can have severe consequences. In 1998, a British doctor suggested a link between the three-in-one vaccine for mumps, measles and rubella, known as MMR, and the onset of the behavioural disorder known as autism. Even though the British government assured parents that the MMR jab was safe, supported by numerous studies that found no link between MMR and autism, many parents decided it was safer not to give their children the jab.

As a consequence, cases of measles and mumps in the UK soared (rubella is still thankfully rare). Measles cases rose from 56 in 1998 to a peak of over 2000 in 2012, while mumps cases rose from 121 in 1998 to a peak of 45,000 in 2005. Not only are measles and mumps highly unpleasant diseases, but measles can also occasionally kill. The number of cases of both measles and mumps has thankfully fallen since, but they remain above the levels of the mid-1990s, before the scare.

Similar anti-vaccination sentiment has led to large rises in measles cases in many other parts of the world, including in countries such as the US where the disease had essentially been eradicated. There were over 644 cases in the US in 2014 – more than in the previous five years combined – and 372 cases in 2018. In the past few years, several European countries, including Italy, France and Germany, have experienced major measles outbreaks.

A highly speculative, and later proven to be unfounded, link between MMR and autism was sufficient to stop parents vaccinating their children against three serious, but eminently preventable, diseases.

Usually, scientists are fighting against System 1 to assure people that a technology is safe, but occasionally they are fighting against System 1 to get people to take a threat more seriously. This is the case with global warming.

Spotlight: A nudge in the right direction

System 1 makes decisions quickly by not thinking about them too deeply. Often this can lead to some rather dodgy decision-making, but it also means this decision-making can be manipulated for the better. This is the idea behind 'nudging', or using knowledge of how System 1 works to steer people's behaviour, usually without them realizing it. Shops have been doing this for years – by placing impulse buys such as chocolate near the tills, for example – but now governments are getting in on the act.

While this may all sound rather Orwellian, governments are usually trying to nudge people to behave in ways that are of benefit to them or society in general, by making the desired behaviour easy, attractive, social and timely. Easy can be achieved by making the desired behaviour the default option, rather than asking people to sign up for it. This works well for tasks with obvious benefits that people nevertheless tend to put off doing, such as agreeing to donate organs after death and enrolling on a pension scheme.

Making a behaviour more attractive for specific groups can often be achieved by simply changing the wording. A study in France on the teaching of technical drawing found that boys did better if the subject was called 'geometry', but girls did equally well or better if it was called 'drawing'. Social nudging takes advantage of a person's desire to confirm to social norms, and to be seen to be conforming. A trial in the US found that telling profligate users of energy how their consumption compared with that of their neighbours prompted them to use less.

Not all this 'nudging' is so altruistic, however, because governments are also using it to get people to pay their taxes on time. One approach taken in Singapore is to print tax bills on the pink paper typically used for debt collection. A slightly more aggressive approach tried in the UK involved rewording a letter sent to non-payers of vehicle taxes. The new version bluntly told them to 'pay your tax or lose your car', sometimes with a photograph of the car in question. This tripled the number of non-payers that paid the tax.

Key ideas

- Humans appear to possess two separate systems of thought, termed 'System 1' and 'System 2'.
- System 1 is essentially intuition or gut feeling; it uses some basic rules of thumb, or heuristics, to make a decision quickly.
- System 2 is all about dispassionately considering the evidence; it's logical and systematic, but it's also slow and takes a lot of mental effort.
- The three main heuristics used by System 1 are the anchoring and adjustment heuristic, the representativeness heuristic and the availability heuristic.
- System 1 also utilizes numerous 'biases' or tendencies, including the tendency to seek out information that corresponds to our existing beliefs and to believe that risk and benefit are inversely related.

Dig deeper

Gardner, Dan, *Risk: The science and politics of fear* (London: Virgin Books, 2009).

Halpern, David, *Inside the Nudge Unit: How small changes can make a big difference* (London: W. H. Allen, 2015).

23

Hot enough for you?

Climate change has always been controversial. Also known as global warming, this is the theory that rising concentrations of carbon dioxide and other gases, mainly produced by our burning of fossil fuels for energy, are causing the Earth to heat up, with potentially dramatic consequences for us and life in general.

Climate change generates more ire and controversy than almost any other area of science. It also seems to be an area where people are forced to choose: you're either a believer in climate change or you're not – there is no middle ground. As a result, the level of antagonism between the two sides has always been high.

One of the reasons for the ire and controversy is that climate change is a slow process and so there has never been a 'smoking gun' that climate scientists can point to as incontrovertible proof that climate change is happening. Nevertheless, over the past decade or so, the growing weight of evidence in support of climate change has become increasingly impossible to ignore: ice sheets melting in the Arctic and Antarctic at unprecedented rates; once-in-a-hundred-year droughts, floods and hurricanes happening increasingly frequently; coral reefs disappearing from warming oceans across the world ... all events that scientists predicted would happen thanks to climate change and all now coming to pass. So, while climate change remains controversial, there's less and less reason why it should be so.

The principles of climate change

Ironically, the basic principles of climate change are, in essence, very simple and uncontroversial. Carbon dioxide is a greenhouse gas, as are methane (also known as natural gas) and nitrous oxide. As we saw in Chapter 17, different molecules absorb and emit electromagnetic radiation at different frequencies; carbon dioxide, methane and nitrous oxide all happen to absorb radiation at infrared frequencies.

Carbon dioxide, methane and nitrous oxide are all present in the atmosphere, albeit at very small concentrations (carbon dioxide currently accounts for around 0.04 per cent of the atmosphere, while methane and nitrous oxide are present at even lower concentrations). The vast majority of the atmosphere consists of nitrogen and oxygen, accounting for 78 per cent and 21 per cent respectively, but neither of these absorbs infrared radiation.

GREENHOUSE GASES

Greenhouse gases are so-called because by absorbing infrared radiation they prevent heat escaping from the Earth, in a similar manner to a greenhouse. This infrared radiation is emitted by the surface of the Earth as it is warmed by sunlight (see Chapter 17). A good example of this kind of surface heating is provided by tarmac, which gets very hot on sunny days.

If the atmosphere consisted of nothing but nitrogen and oxygen, then this infrared radiation would simply escape back into space. Instead, much of it is absorbed by greenhouse gas molecules, causing them to move more rapidly and emit more infrared radiation back towards the ground. In this way, carbon dioxide, methane and nitrous oxide act as a kind of thermal blanket over the Earth, preventing the surface from cooling down.

Scientists have calculated that without these greenhouse gases in the atmosphere, the average temperature at the surface of the Earth would be a chilly –18°C, meaning that greenhouse gases are essential for making the Earth habitable.

A quick look at Mars will confirm this point. Carbon dioxide is the main constituent of the Martian atmosphere (accounting for around 95 per cent), but this is more than offset by the fact that its atmosphere is over 100 times thinner than ours and so it traps far less heat. As a result, the average temperature at the surface of Mars is –65°C.

But you can have too much of a good thing. Take Venus: carbon dioxide is the main constituent of its atmosphere as well, but its atmosphere is 100 times thicker than the Earth's. As a result, the temperature at the surface of Venus is a furnace-like 460°C. So, broadly speaking, the more greenhouse gases you have in your atmosphere, the more heat is trapped near the surface.

INCREASED LEVELS OF CARBON DIOXIDE

And the concentration of greenhouse gases, particularly carbon dioxide, in the Earth's atmosphere is increasing. Again, this is an uncontroversial finding. Measurements of atmospheric carbon dioxide levels have been made since the 1950s and show a steady increase, from 317 parts per million (ppm) in 1959 to 405 ppm in 2017. Also uncontroversial is attributing this increase to humankind's use of carbon-rich fossil fuels such as oil, coal and gas, which produce carbon dioxide when burned.

Additional evidence is provided by studies on bubbles of air trapped in ice cores extracted from the Arctic and Antarctic. These provide an atmospheric record stretching back hundreds of thousands of years and indicate the carbon dioxide concentrations have been rising for 250 years, ever since the dawn of the industrial revolution. Although the ice cores also show that carbon dioxide concentrations naturally fluctuate, these concentrations are now the highest they've been for at least 800,000 years.

Controversy over rising air temperatures

The controversy starts heating up with the assertion that average surface air temperatures are increasing and that rising carbon dioxide levels are to blame. For although, broadly speaking, more carbon dioxide means more trapped heat, carbon dioxide is just one of the many factors that influence surface air temperature.

The precise relationship between temperature and carbon dioxide is highly complex and one that scientists don't yet fully understand.

As well as providing details of historical carbon dioxide levels, ice cores can also provide information about historical temperatures, because different isotopes of oxygen are laid down in the ice at different rates depending on the temperature. Comparing this temperature record with the carbon dioxide record reveals that the two do tend to fluctuate in step. But ice cores can't reveal the precise relationship: whether the temperature fluctuates in response to fluctuating carbon dioxide levels, or vice versa; or, more likely, whether there's a complex interaction between the two.

This temperature record is also fairly coarse, which is not surprising considering that it covers hundreds of thousands of years. For more detailed records of recent history, scientists need to turn to other natural sources, such as tree rings. These mark not only a tree's age but also the temperatures it has been exposed to over its lifetime, which influence the thickness of the rings.

Using tree rings from various ancient trees, scientists have built up a record of average global temperatures over the past 1,000 years. Fluctuations are once again the name of the game, but always around a general mean.

But that changed about 150 years ago, when humankind first started taking regular measurements of the air temperature. These measurements, which are now taken at thousands of weather stations around the world and by satellites, show generally rising temperatures (although the temperature was flat from the mid-1940s to the mid-1970s). Over the past 150 years, the world's average surface air temperature has warmed by around 1°C.

To many critics this sounds suspiciously convenient, especially as this recent warming doesn't appear to show up in tree rings. They think this warming may have more to do with the fact that, over the years, towns and cities, which tend to be hotter than the surrounding countryside, have often built up around formerly isolated weather stations. However, climate change scientists argue that they have taken this urban warming into consideration – by, for instance, omitting readings from stations in the largest urban areas – when putting together their records of average temperature.

Even if the surface air temperature is increasing, rising carbon dioxide levels are not necessarily the sole culprit. For a start, as we have seen, carbon dioxide is not the only greenhouse gas. In fact, carbon dioxide is actually a less potent greenhouse gas than either methane or nitrous oxide, but it matters more because there is more of it in the atmosphere. Overall, it has been estimated that methane contributes about 24 per cent to global warming, while carbon dioxide contributes 70 per cent.

The effect of water vapour

But an even more important greenhouse gas than carbon dioxide, because it's much more prevalent in the atmosphere, is water vapour. Critics therefore argue that a small rise in carbon dioxide levels is relatively unimportant, because water vapour has a much larger influence on surface air temperatures. But things are not that simple.

For a start, water vapour would amplify any warming produced by rising carbon dioxide levels, because a warmer surface temperature causes more evaporation from the oceans and thus more water vapour. On the other hand, as we saw in Chapter 15, water vapour quickly condenses to form clouds, and clouds, especially white clouds, reflect incoming solar radiation, thereby helping to cool the surface. But then again, clouds can also trap some of the infrared radiation emitted by the ground.

So, water vapour has conflicting effects on temperature, although scientists think that, overall, more water vapour probably leads to higher temperatures. Carbon dioxide, on the other hand, only leads to warming, at least near the Earth's surface.

Solar radiation

With similar reasoning, critics also argue that the amount of radiation the Earth receives from the Sun, which varies on numerous timescales, has a much greater effect on surface air temperatures than carbon dioxide levels. If the Earth is warming, they say, then this must mainly be due to it receiving more energy from the Sun.

But this argument is contradicted by observations of the Sun, which indicate little change in incoming solar energy over the past few decades, and also by the findings of computer-based climate models. By chopping up the atmosphere and the Earth's surface into three-dimensional cells and then simulating various physical processes in those cells, these models attempt to replicate the workings of the climate.

Even though such models can only provide a very general approximation of the real climate, they have proved pretty accurate at replicating the warming seen over the past 100 years. They can only do this, though, by incorporating rising carbon dioxide levels. According to the models, fluctuations in solar radiation cannot on their own account for the observed warming.

Changes in the future

The models are much less certain about what will happen in the future, although they all indicate further warming. This lack of certainty is mainly down to there still being much that we don't understand about how the climate works. We are even unclear about what happens to all the heat trapped at the surface by carbon dioxide.

Only a small proportion of the heat remains in the lower atmosphere and contributes directly to global warming; the vast majority is absorbed by the oceans, causing them to warm up, or goes towards melting glaciers and ice caps at the poles. Scientists had thought that this heat was mainly being stored by surface waters, but recent research indicates that deeper ocean waters, below 700 metres, may be storing more heat than realized. How the heat is working its way down there is not yet known.

Still, global surface temperatures remain on an upward trend. Despite a slowdown between 1998 and 2013, which may be due to more heat being stored in deep ocean waters or merely an artefact produced by inconsistent measurements, 2015 to 2018 were the four hottest years on record, with 2016 taking the title.

As this trend continues, sea levels will start to rise, causing many coastal areas to disappear under the waves. This will be due to both warming oceans, because water expands as it warms, and melting sea ice and glaciers. Indeed, this process is already happening, with satellite measurements indicating that the seas are currently rising by over 3 mm a year.

Rising ocean temperatures and meting sea ice may also change the ocean circulation system (see Chapter 14), potentially slowing down or even switching off the thermohaline circulation (as happened in the 2004 disaster film *The Day after Tomorrow*, with dramatic but scientifically unlikely consequences). This would have the rather ironic effect of making Western Europe much colder, although there's no sign of it happening yet.

Almost everywhere will become warmer, although this warming will probably be more dramatic at night and in the winter than during the day and in the summer. The warmer temperatures will increase evaporation of water from the oceans, making it rain harder when it does rain, but they will also increase evaporation from many inland areas, making droughts more intense. Again, this is already beginning to happen.

Many plants and animals will suffer, as their habitats change. It has been predicted that up to 37 per cent of plant and animal species could face extinction by 2050 if carbon dioxide levels keep on rising. It's true that more carbon dioxide in the atmosphere could cause plants to grow faster, but this will likely be offset by higher temperatures. As a result, our food supply could suffer: the yields of our most important food crops fall dramatically when temperatures are much above 30°C.

Despite the continuing controversy, virtually all the world's major scientific societies, including Britain's Royal Society and the US National Academy of Sciences, accept that global warming is a major problem and that rising levels of carbon dioxide, produced by the burning of fossil fuels, are mostly to blame. They are thus trying to move the debate away from whether global warming is happening and towards what to do about it (see Spotlight).

Spotlight: Finding reverse gear

Most countries now accept the need to reduce their carbon dioxide emissions, even if global attempts to set binding targets have had mixed fortunes. Ways to reduce emissions can include replacing coal-fired power stations with renewable forms of energy, swapping petrol-powered cars for electric vehicles and making all our technologies as energy-efficient as possible. All of which are being done, but much too slowly.

To prevent the worst effects of climate change, the Intergovernmental Panel on Climate Change, a body set up to advise governments, has recommended keeping the average global increase in surface air temperature to less than 2°C above pre-industrial levels, and preferably to less than 1.5°C. On current emission trends, however, we are comfortably going to overshoot the 2°C target, let alone the 1.5°C target.

This is spurring scientists to think of more direct means for slowing the warming. One way is to capture the carbon dioxide emitted by power plants and industry before it reaches the atmosphere, using some form of absorbent material. The carbon dioxide can then be stored safely underground or even converted into useful products, such as fuels or plastics.

Alternatively, carbon dioxide already in the atmosphere could be actively removed. The simplest way of doing this is to grow more plants, especially trees, because they soak up carbon dioxide as they grow, but this is a slow process and requires lots of land. Various technological approaches are also being developed, which tend to involve pumping air through a material that absorbs or reacts with carbon dioxide.

Another way is to ignore the carbon dioxide, but instead reduce the amount of sunlight reaching the ground. One way to do this would be to release loads of particles into the upper atmosphere, where they would reflect some of the incoming sunlight back into space. Scientists are fairly confident this would work, at least temporarily, because this is exactly what happens when large volcanic eruptions blast vast quantities of ash into the atmosphere.

Key ideas

- Global warming is the theory that rising concentrations of carbon dioxide and other gases, mainly produced by our burning of fossil fuels for energy, are causing the Earth to heat up.
- Greenhouse gases are so-called because by absorbing infrared radiation they prevent heat escaping from the Earth, in a similar manner to a greenhouse.
- Measurements show that the concentration of greenhouse gases, particularly carbon dioxide, in the Earth's atmosphere is increasing.
- The controversy starts heating up with the assertion that average surface air temperatures are increasing and that rising carbon dioxide levels are to blame.
- Global surface air temperatures are on an upward trend: 2015 to 2018 were the four hottest years on record, with 2016 taking the title.

Dig deeper

Romm, Joseph, *Climate Change: What everyone needs to know* (New York: Oxford University Press, 2018).

Wallace-Wells, David, *The Uninhabitable Earth: A story of the future* (New York: Random House, 2019).

Apocalypse now

As global warming amply demonstrates, humankind's technological prowess has reached a stage where it can now threaten the entire planet. And it's not just global warming: we can now destroy our world in a whole variety of weird and wonderful ways, from nuclear war to genetically modified disease to black holes. If we're really unlucky, we may even take the rest of the universe with us.

But before exploring the various options for global annihilation, we need to make a distinction between two main types. In the first, and less catastrophic, type, it is our modern civilization that is destroyed rather than the Earth. This may involve humankind merely regressing to a less technologically advanced state following some tragedy. But it could also involve humankind being wiped off the face of the Earth, probably along with many other species.

The Earth as a whole would survive this first type of annihilation. Life in some form or other would also probably survive, thriving again after a certain period of time, as happens after mass extinctions (see Chapter 5). Humankind would no longer be around, though. The second, more catastrophic, type of annihilation involves the physical destruction of the Earth, in which case obviously nothing would survive.

Nuclear war

For many years, the most likely cause of the first type of global annihilation was nuclear war. It has been estimated that in the mid-1980s the US and the Soviet Union (now Russia) had around 65,000 nuclear warheads between them, equivalent to the explosive power of 3 tonnes of TNT for every person on the planet.

The immense destructive power of nuclear weapons is down to the fact that, per kilogram, a nuclear reaction is a million times more efficient at generating energy than a chemical explosion. In an atomic bomb, the explosion is caused by the same kind of fission reactions that take place in nuclear power stations (see Chapter 16), whereby flying neutrons split uranium atoms. The difference is that in an atomic bomb these reactions run out of control, releasing loads of heat and atomic particles in a massive explosion equivalent to over 10,000 tonnes of TNT.

THE HYDROGEN BOMB

In more advanced nuclear weapons, such as the hydrogen bomb, the explosion is caused by both fission and fusion reactions. Fusion reactions are the opposite of fission reactions, with atomic nuclei fusing together (as happens in the cores of stars; see Chapter 2), but they still release lots of energy in the form of heat and atomic particles. By combining fission and fusion reactions, with the fusion reactions involving isotopes of hydrogen, a single hydrogen bomb can produce an explosion that is equivalent to millions of tonnes of TNT (hence the term megatons).

But it is not just the size of the initial explosion that makes nuclear weapons so horrifying. The force of the explosion from a modern hydrogen bomb, together with the resultant massive fires, would send huge quantities of material up into the atmosphere, in the form of smoke and particles of rock and soil. In all likelihood, this material could stay up there for years, blocking sunlight and ushering in a nuclear winter that would make life tough for those that survive. In particular, the reduced sunlight and lower temperatures would decimate our ability to grow crops.

IONIZING RADIATION

A nuclear blast also releases high levels of ionizing radiation, in the form of massive numbers of highly energetic atomic nuclei and subatomic particles that can damage chemical bonds. Spreading over a large area, this ionising radiation would cause both immediate and long-term damage to survivors, including burns and genetic damage that could lead to high rates of diseases such as cancer. It would also affect future generations by greatly increasing the risk of birth defects. A global nuclear war would clearly kill off our modern civilization, if not much of humanity and many other species.

Nuclear threats

With the end of the Cold War, this nightmare scenario has become much less likely, especially as both sides have been steadily reducing their nuclear stockpiles. But, although the threat of global annihilation via nuclear weapons seems to have passed, the danger that someone somewhere will use a nuclear weapon is still very much with us.

That could be unstable, erratic regimes in countries such as North Korea, which has developed and tested nuclear weapons, and Iran, which has tried to develop nuclear weapons in the past and may well again in the future. It could also be terrorist groups that have obtained a bomb from these unstable regimes or from old Soviet stockpiles.

Smaller versions of the Cold War are also still being played out. Both India and Pakistan have nuclear weapons and are currently engaged in a stand-off over the disputed territory of Kashmir. If Iran does ever develop a nuclear bomb, then it will likely engage in a stand-off with Israel, which many believe already has nuclear weapons. A nuclear conflict between these countries would not result in global annihilation, but would still kill millions.

Chemical and biological attacks

Large numbers could also die from attacks with chemical and biological weapons. Such weapons are now banned under the Chemical Weapons Convention and the Biological and Toxins Weapons Convention, but that obviously doesn't stop terrorists and deranged cults from using them.

In fact, they already have. In 1995, members of a Japanese cult called Aum Shinrikyo released the nerve gas sarin in the Tokyo subway, killing 12 people. But nasty as chemical weapons are, biological weapons, which are based on disease-causing pathogens, pose a much greater threat. A chemical attack will only affect people in the immediate vicinity; a biological attack, on the other hand, has the potential to spread far and wide. This is both because an infectious biological pathogen can spread from person to person and because it may be several days before the authorities work out that anything is wrong.

DESIGNER PATHOGENS

Natural pathogens are deadly enough. For instance, smallpox is highly infectious and kills around a third of those infected; fortunately, widespread vaccination has all but wiped out this former scourge of humanity (see Chapter 9). But a growing danger comes from the possibility that someone will use modern genetic and synthetic biology techniques (see Chapter 20) to create a novel, designer pathogen that is more infectious and deadly than anything nature has yet conjured up.

Imagine a version of the Ebola virus that could be transmitted through the air, rather than through direct contact with bodily fluids, or a version of the flu virus that kills a third or more of those infected. If released in a modern city, such an infection would quickly sweep the world, as happened with swine flu in 2009, despite all efforts to contain it. It could kill hundreds of millions.

DISASTROUS CONSEQUENCES

But a nuclear or biological attack would not need to kill millions to threaten our modern civilization; the disruption caused by a smaller attack could do that. Indeed, an attack in which no one died could do it. The point is that in many ways our modern, technological civilization is quite fragile, heavily reliant on things such as energy supplies, food distribution and communication technologies.

A small nuclear or biological attack would disrupt all of these things, resulting in widespread electricity blackouts, and associated food, water and fuel shortages. A cyber attack that took out large swathes of our global computer network could do much the same thing, with no immediate loss of life – as could something as seemingly innocuous as a solar flare (see Spotlight). In an instant, modern societies would regress hundreds of years, a setback from which it would take a long time to recover, and during which many could die from disease and any collapse in law and order.

Spotlight: Flare for destruction

In addition to light and heat, the Sun is continually releasing streams of charged subatomic particles, known as solar wind, into space. Occasionally, it also releases huge, powerful bursts of these particles, known as solar flares or coronal mass ejections (CMEs), which have the potential to cause substantial damage to electricity networks if they strike the Earth. Damage that could shut them down for months or years.

Actually, solar flares and CMEs are not quite the same thing. Solar flares are sudden increases in the brightness of a specific region of the Sun, which are usually, but not always, caused by CMEs. These CMEs can vary quite a bit in size and usually blast off harmlessly into space, but sometimes they are fired towards the Earth.

In 1989, a not-particularly-large CME took out the electricity network of Quebec in Canada for nine hours. In 2012, a much larger CME just missed the Earth; had this CME hit the US, it could have destroyed a quarter of the country's high-voltage transformers. The most powerful CME in modern history hit the Earth in 1859, generating auroras that were seen in the tropics and damaging the nascent electric telegraph system. If such a CME were to the hit the Earth today, it would fry the electricity networks of many countries around the world and take out a host of satellites, plunging the world into darkness, both physical and metaphorical.

There are ways to protect high-voltage transformers and other components of the electricity network from the effects of CMEs, such as by fitting surge protectors, but this would be costly and so has not been widely done. Several satellites launched to study the Sun could also provide early warnings of an incoming CME, allowing the most vulnerable equipment in the network to be switched off before it hits.

Long-term damage to the Earth

Still, none of these various apocalypses would cause any major long-term damage to the Earth. Even setting off the tens of thousands of nuclear weapons held by the US and the Soviet Union at the height of the Cold War wouldn't have caused it too much trouble. The asteroid blamed for killing off the dinosaurs 65 million years ago released as much energy as 1 million hydrogen bombs. Apart from a bit of a dent, in the form of a 200-km-wide crater straddling the coast of Mexico, the Earth emerged pretty much unscathed.

So surely there's no way that humankind could possibly destroy the Earth. Well, actually there is, and the potential danger comes from a rather surprising source. As we saw in Chapter 1, in order to determine the fundamental structure and forces of the universe, physicists are slamming subatomic particles together at close to the speed of light in particle accelerators such as the Large Hadron Collider (LHC).

Spotlight: Assessing the apocalypse

How can you assess apocalypses dispassionately? How can you compare the severity of different apocalypses and determine what steps should be taken to prevent an apocalypse from taking place? Well, it turns out there is a simple method for assessing the severity of any threat that could result in human casualties, be it a hurricane, asteroid impact or nuclear war.

It involves multiplying the probability that the threat will happen by the magnitude of the threat, usually in terms of the number of likely casualties. Using this method, a threat that only kills a few people but is quite likely to happen is rated as severe as a threat that is very unlikely to happen but would kill many more people.

This explains why a number of efforts are now under way to catalogue all the asteroids and comets that could cross Earth's orbit, known as near-Earth objects. The kind of asteroid impact that wiped out the dinosaurs may be very rare, occurring every 100 million years on average, but the consequences would clearly be apocalyptic.

BLACK HOLES, STRANGELETS AND OTHER CALAMITIES

For a brief moment, these collisions generate an immense amount of energy, replicating the conditions at the birth of the universe. The concern is that these high-energy collisions could potentially trigger the destruction of the Earth. There are a number of ways this could theoretically happen.

The high-energy collisions could create a microscopic black hole that proceeds to gobble up the Earth (see Chapter 2). Alternatively, they could cause the quarks that ordinarily form protons and neutrons to reassemble into a strange form of matter known as a strangelet. This could then proceed to convert all other matter, transforming the Earth into an immensely dense sphere just 100 m across. Or the collisions could convert the whole of the universe into a new stable state, known as a vacuum bubble, in which atoms could not exist. Finally, they could produce particles known as magnetic monopoles, which some theories suggest can cause protons to decay, in which case atoms would simply melt away.

Although all these possible calamities are more or less consistent with current theories, they are highly speculative, and most scientists dismiss them. Also, similar high-energy collisions occur naturally all the time when cosmic rays (see Chapter 2) hit atoms in the Earth's atmosphere and these collisions haven't yet destroyed the Earth.

All the same, if you suddenly find yourself converted into a hyper-dense lump of strange matter then you'll know who to blame.

Key ideas

- A nuclear reaction is a million times more efficient at generating energy than a chemical explosion.
- Modern genetic and synthetic biology techniques could be used to create a designer pathogen that is more infectious and deadly than anything nature has yet conjured up.
- The disruption caused by a small nuclear or cyber attack, or a solar flare, could threaten our modern civilization.
- The asteroid blamed for killing off the dinosaurs 65 million years ago released as much energy as 1 million hydrogen bombs, but the Earth emerged pretty much unscathed.
- There are several ways in which the high-energy collisions taking place in particle accelerators such as the Large Hadron Collider could theoretically trigger the destruction of the Earth.

Dig deeper

Dartnell, Lewis, *The Knowledge: How to rebuild our world after an apocalypse* (London: Vintage, 2015).

Rees, Martin, *Our Final Century: Will civilization survive the twenty-first century?* (London: Arrow Books, 2004).

25

Know your limits

The reason why scientists consider science to be the most accurate and reliable repository of knowledge about the universe, compared to religious revelation and philosophical musing, is all down to the scientific method. In essence, the scientific method is very simple: a scientist makes certain observations about the universe, develops a hypothesis about how the universe works based on those observations and then conducts experiments to test that hypothesis.

Simple it may be, but this method has generated some pretty fundamental insights (many of which are described in this book). Furthermore, as if to prove the accuracy of these insights, it has allowed us to develop all the technology that we enjoy today.

Now, in practice, the method is a bit more complex than just described. For a start, it's an iterative process: the results of the experiments performed to test a particular hypothesis often lead to modifications of the original hypothesis or the formation of a new hypothesis, leading to even more experiments. Furthermore, at some point all these hypotheses need to be combined into an overarching theory that accurately explains why the universe is as it is. Ideally, this theory should also make predictions that can be tested by even more experiments.

Setting restrictions on the universe

Certain restrictions also need to be placed on the universe for the scientific method to work properly. The scientific method is essentially all about discovering cause-and-effect relationships: in an experiment, a scientist is looking for a specific effect from a specific action. To do this, he or she (just to keep things simple, we'll assume 'he' from now on) needs to make sure that nothing else can influence the outcome.

For example, if a scientist wants to study the effect of rising temperatures on a crop plant such as wheat, he'll need to keep a tight control on all the other factors that could affect the plant. These include: humidity; nutrient and water levels in the soil; sunlight intensity; and carbon dioxide levels. Otherwise the scientist won't be able to tell whether any recorded changes to the plant were due to rising temperatures or to one or more of these other factors.

But the real world isn't like this. In the real world, plants are affected by all the factors simultaneously and in combination, such that it is the interaction between these factors that truly determine the effect on the plant. Studying all of these factors individually is known as reductionism and it has allowed scientists to work out in some detail how individual parts of the universe work, such as the various biological molecules within a cell. But scientists know far less about how these different parts interact with each other and mesh together.

The scientific method also assumes that the scientist is independent of the process or relationship he is studying, meaning that he doesn't influence it in any way. The effects of rising temperature on a wheat plant would be just the same whether the scientist was recording them or not.

SCIENCE VERSUS RELIGION AND PHILOSOPHY

Despite these restrictions and assumptions, the scientific method has so far proved superior to any other method for accumulating knowledge about the universe. It trumps religious revelation because its theories can be tested by experiment; those that come up short are soon replaced or discarded, unlike religious tenets. It trumps philosophical musing because its theories are constructed from experimental evidence, rather than from sitting around thinking about things.

Beyond the scientific method

Recently, however, science has begun to outgrow the scientific method, at least in its classic incarnation. There are a number of reasons for this. In part, it's because all the fundamental discoveries have already been made; in part, it's because modern technology, especially computers, offers new ways to study the universe; and, in part, it's because modern science is now exploring areas where the assumptions of the classic scientific method no longer apply.

Now, it's obviously incredibly presumptuous to assert that science has already made every fundamental discovery about how the universe works. It's also plainly untrue. To mark its 125th anniversary in 2005, the scientific journal *Science* produced a list of the 25 big, unanswered questions facing science, which were picked out from an initial list of 125. These questions included things like: What is the universe made of? What is the biological basis of consciousness? Can the laws of physics be unified? How and where did life on Earth arise? Are we alone in the universe? All of these questions still remain unanswered, and although scientists have made progress on some of the other 20, none has been answered fully.

So, there are obviously many more important discoveries to be made, but the nature of those discoveries does appear to be changing. Most of these 25 questions can be answered only by studying complex systems as a whole, rather than studying different aspects of those systems in isolation.

It's not that all the major scientific discoveries have been made, but that all the major discoveries that can be made using a reductionist approach may well have been made. We know how elements are forged in stars; we've worked out the structure of DNA and proteins; and we understand why some materials conduct electricity better than others. And this understanding came from countless experiments studying specific aspects of each phenomenon.

That approach just won't suffice, however, for determining the workings of complex systems such as human consciousness and the climate, where the interaction between different elements is more important than the elements themselves. Understanding how a single neuron works can't explain human consciousness and understanding how carbon dioxide absorbs infrared radiation can't explain the climate.

The use of computers in science

This is where computers come into their own, because one way for scientists to probe complex systems is to develop computerized models of them. As with the climate models mentioned in Chapter 23, these models represent a simplified version of the real system, replicating known physical or biological processes and the interactions between them across the system. The ability to produce such computerized models has already led to the rise of completely novel scientific disciplines. One example is systems biology, which aims to understand the full suite of complex interactions taking place between the many biological molecules, including genes and proteins, in cells.

Indeed, computers are taking over more and more of the scientific discovery process. Many scientific studies are now entirely automated, from protein analysis to astronomical surveys to material synthesis, and generate huge amounts of data. So much, in fact, that scientists also need computers to analyse the data and highlight any trends and patterns.

Obviously, the potential problem with this growing reliance on computers is that scientists will understand less and less of what the computers are doing and the results they are producing (see Spotlight). Already, computer models can be a bit of a black box, accurately replicating certain aspects of a complex system without scientists really understanding how they do it. But still, computers and computer models are probably going to be essential tools for studying complex systems, seeing as reductionism on its own is not going to cut it.

Spotlight: Software slip-ups

Computers have been an undeniable boon for scientists, especially as they move from studying simple cause-and-effect relationships to complex systems. Such studies rely on using computers to collect, store, process and analyse huge reams of data, implementing complex mathematical techniques to tease out underlying trends and patterns. But as the complexity of these mathematical techniques increases, so does the possibility they may go awry, finding patterns where there are none.

Concerns of this type have been raised about functional magnetic resonance imaging (MRI), in which the MRI scanners usually used for medical investigations are instead used to probe brain function. This involves MRI scanners monitoring blood flow in the brain by distinguishing between oxygenated and deoxygenated states of haemoglobin, the molecule in blood cells that carries oxygen. The idea is that blood flow will be greater to active areas of the brain, allowing functional MRI to identify regions of the brain responsible for specific functions, from memory to speech.

To identify these active areas, scientists need to compare the blood flow in subjects performing some specific task with the blood flow in controls who aren't doing anything. This requires a lot of computer-based data processing, and a study in 2016 revealed that processing could be fallible. In this study, scientists took control data from several old functional MRI studies, randomly split this data into two groups and then compared them with software packages commonly used to analyse functional MRI data. They did this numerous times, varying the make-up of the two groups each time.

The software packages shouldn't have found any consistent differences between the two groups, because they were both made up of control data, but the software ended up identifying differences up to 70 per cent of the time. This suggests that many of the findings of actual functional MRI studies could be erroneous, falsely identifying regions of the brain as responsible for a specific function. Because similarly complex data processing is used by many other scientific disciplines, some of their findings could be equally suspect, reinforcing the importance of replicating studies (see Chapter 21).

Quantum theory

If complex systems are difficult to understand, then the quantum world is completely impossible. As the late US physicist Richard Feynman famously put it: 'If anyone claims to know what the quantum theory is all about, they haven't understood it.' Quantum theory, also known as quantum mechanics, explains the behaviour of matter and energy at the level of individual atoms and below, and what it says is that at these tiny scales all the certainties of our normal world break down.

WAVE–PARTICLE DUALITY

In Chapter 17, we learned that visible light is a form of electromagnetic radiation and thus acts like a wave, which is a demonstrable, experimentally proven fact. But in some circumstances visible light also acts like a particle known as a photon, which is also a demonstrable, experimentally proven fact. It sounds nonsensical, but in some cases light acts like a wave and in some cases it acts like a particle. This is known as wave–particle duality.

Even more astonishingly, it's not just light that behaves like this, but also electrons, protons, atoms and even molecules. At times, they act like a wave and at other times like particles. This helps to explain another troubling aspect of the quantum world: that, unlike in the normal world, we can't know both the position and the momentum (or speed) of a quantum entity like an electron. In fact, the more we know about the momentum of an electron, the less we know about its position and vice-versa. It actually turns out that we can know the momentum of an electron when it's behaving like a wave and know the position when it's behaving like a particle.

Furthermore, this propensity to behave like both a wave and a particle extends to other properties, such that a single subatomic particle can possess two conflicting properties at the same time. For instance, an electron possesses a property known as spin, which is related to magnetism and can be either up or down. But it is only up or down when the property is actually measured, before that the spin of the electron is both up and down at the same time. This ability to exist in what is known as a 'superposition' of states is fundamental to the development of quantum computers (see Chapter 27).

It also means that scientists can't be objective and independent when studying the quantum world. In their undisturbed state, subatomic particles exist in a superposition of states. It's only when scientists actually probe these particles that they adopt a defined state, such as a specific position or spin. Thus, scientists can't study the quantum world without affecting it, meaning that the classic scientific method by necessity falls down.

Because of this, there is still much about the quantum world that scientists don't understand. This includes how the indeterminacy of the quantum world transforms into the certainty of our normal world, where objects can't possess conflicting properties and where we can know both their position and momentum. It may be that the quantum world is just too different for us ever to understand it fully (see Spotlight).

Nevertheless, scientists will never stop trying to understand it or to explore the many other questions that remain unanswered, even if this requires going far beyond the classic scientific method. For science, the future holds a myriad of exciting possibilities, even though in many cases science fiction got there first.

Spotlight: How much can we understand?

The bizarre and confusing nature of the quantum world raises the possibility that there are some things humans will just never understand. Our large brains evolved to help us survive in groups on African savannahs, not to uncover the intricacies of the universe. It may be that we just don't have the mental capacities to probe to the very depths of reality; after all, we don't expect fish to understand algebra.

Alternatively, it may be that there are various different ways to understand the universe and that our brains are only capable of perceiving one of those ways. Alien life forms with different brains may perceive the universe in completely different, but still consistent, ways. Or perhaps we are imposing structure on to the universe; protons and electrons may exist solely in our scientific instruments and theories. In that case, we are not discovering scientific knowledge but inventing it.

Key ideas

- The scientific method involves making certain observations about the universe, developing a hypothesis about how the universe works based on those observations and then conducting experiments to test that hypothesis.
- Studying factors individually is known as reductionism and it has allowed scientists to work out in some detail how individual parts of the universe work.
- A reductionist approach won't suffice for determining the workings of complex systems such as human consciousness and the climate.
- Computers and computer models are probably going to be essential tools for studying complex systems.
- Quantum theory explains the behaviour of matter and energy at the level of individual atoms and below, and what it says is that at these tiny scales all the certainties of our normal world break down.

Dig deeper

Al-Khalili, Jim, *Quantum: A guide for the perplexed* (London: Weidenfeld & Nicolson, 2012).

Horgan, John, *End of Science: Facing the limits of knowledge in the twilight of the scientific age* (New York: Basic Books, 2015).

Part Six

Science of the future

Back to the future

By now, we should have flying cars, moon bases and androids. Instead, we've got wireless speakers, transparent spray-on sun lotion and smartphones; all impressive technologies, but very different from what was predicted.

'Prediction is very difficult, especially about the future' is an aphorism that has been attributed to various different people, from Nobel Prize-winning physicist Niels Bohr to the US baseball player and coach Yogi Berra. But it's true enough, and never more so than for predictions of future technologies.

One of the reasons why it's so difficult to make accurate technological predictions is that just because a technology seems theoretically possible doesn't mean it will ever be developed or become widely adopted.

Incorrect predictions

VIDEOPHONES

Videophones have been on the cards ever since the development of the telephone over 100 years ago, with the US communications company AT&T developing a working videophone in the 1960s. But only with the rise of the internet and services such as Skype and FaceTime are we now able to see who we're talking to.

AT&T's videophone didn't catch on because the system was fairly expensive, and most people don't need to see the person on the other end of the line. The modern, internet version of the videophone doesn't cost any more than a standard phone call (indeed, it often costs less) and so people are much more inclined to use it.

Economics is perhaps the single most important factor in whether or not a new technology becomes a success, with an expensive new technology having to offer substantial benefits over and above existing technologies. If intriguing but not essential, most people will simply wait for the technology to become cheaper, which may of course never happen.

ELECTRIC CARS

But it is finally starting to happen with electric cars, at least those powered by a rechargeable battery. Despite being heralded for many years, electric cars are only now starting to challenge the internal combustion engine. They still don't have the driving range of conventional cars and the fuelling infrastructure is rather patchy, although charging points do now seem to be springing up all over the place. But thanks to the high price of oil-based fuels in many countries, concerns over global warming and polluting exhaust emissions, and the drive of pioneers such as Tesla's Elon Musk, the momentum does finally seem to be shifting towards electric cars.

Whether this momentum will extend to electric cars powered by fuel cells remains to be seen. Fuel cells work by breaking down a fuel such as hydrogen or methane to release electrons that are sent round an external circuit, thereby generating electricity. They have also been around for many years, being used to generate electricity on NASA spacecraft such as the Space Shuttle, and could potentially allow electric cars to travel for longer distances than is currently possible with batteries. But they are expensive, and many technological challenges, such as finding efficient ways to store the hydrogen or methane, still need to be resolved before they can find their way into cars.

SPACE EXPLORATION

It's not just consumer technologies that can fall foul of economic reality. In 1967, *The Futurist*, a US magazine devoted to thinking about the future, asked a selection of the great and good to offer predictions for what would happen in the following decades. Reviewing the accuracy of 34 of these predictions in 1997, the editor of *The Futurist* concluded that 23 were 'hits' and 11 were 'misses'. Most of the misses could be blamed on a single factor: the slowdown in manned space exploration.

Following the first Moon landings on 21 July 1969, it was assumed that by the end of the twentieth century the first permanent bases would be established on the Moon and maybe even Mars. That this didn't happen was mainly down to the huge cost of sending people to the Moon. This cost was deemed worth paying if it meant getting there before the Soviets, but there really wasn't any other rational basis for doing it.

If huge quantities of oil or (more realistically) valuable minerals had been found on the Moon, things might have been different. But as it stands, there is no compelling reason to send a person to the Moon or to establish a base there. As many commentators have pointed out, there are far more important things for governments to spend their money on.

FLYING CARS

Other predicted technologies have proved simply unfeasible or impractical. At the end of the nineteenth century, flying cars seemed to be everywhere, especially in France for some reason. They could be found on French cigarette cards displaying visions of the future, surrounding the Eiffel Tower in a sketch of what life would be like in 1940 and in a print showing people leaving the Parisian opera in 2000.

Once the internal combustion engine had been invented in the late nineteenth century, it seemed only a matter of time before motor cars took to the air. But obviously there's no easy mechanism for doing that. The production of the first modern helicopters in the 1920s led to a slightly more realistic vision of personal helicopters, but that never really happened either. The sad truth is that it's simply much easier, cheaper, safer and more convenient for large numbers of vehicles to travel on the ground, although the recent rise of unmanned aerial vehicles (UAVs), or drones, is beginning to test that idea.

Why do we have rockets?

It's also not just that many predicted technologies never arrive; other important technologies arrive with no warning at all. In 1937, the US National Academy of Sciences organized a study aimed at predicting future technological breakthroughs. Although the study scored some successes, it completely failed to predict some of the most ground-breaking future technologies, including computers, jet aircraft, antibiotics, rockets and nuclear energy.

Spotlight: Where did all these computers come from?

One technology that seemed to catch everyone by surprise was the computer. As mentioned above, a 1937 study of future breakthroughs by the US National Academy of Sciences entirely failed to predict them. Although science-fiction stories of the 1930s were full of robots, spaceships and ray guns, there was hardly any mention of computers.

Even after computers became reality in the late 1940s, no one foresaw how small or ubiquitous they would become, even those developing them. In 1949, the US magazine *Popular Mechanics* would only go so far as to predict that 'Computers in the future may weigh no more than 1.5 tons'.

In the 1940s, the founder of IBM, Thomas J. Watson, allegedly stated that he didn't see the need for any more than a handful of computers in the whole of the US. Even in the 1970s, the chairman of the Digital Equipment Corporation was declaring that 'there is no reason for any individual to have a computer in their home'.

THE EFFECT OF THE SECOND WORLD WAR

But this shouldn't come as too much of a surprise, as it illustrates another major difficulty with making technological predictions: sometimes a single, unforeseen event transforms the entire technological landscape. This was definitely the case with the Second World War, which started a few years after the study and provided the impetus for the development of all these unpredicted technologies.

Computers partly emerged from code-breaking activities; nuclear energy emerged from the atom bomb; and the development of rockets and jet planes were boosted by the war. Although the first antibiotic, penicillin, was discovered in the 1920s, it was in treating casualties during the war that antibiotics really demonstrated their worth. The disruption and altered priorities caused by the war set scientific and technological research off in completely new and unpredicted directions, ultimately leading to our modern world.

Predictions in science fiction

Of course, many proposed future technologies are not offered as specific predictions, but simply as elements in a story. Even before science fiction was a recognized genre, authors have been imagining futuristic technologies, versions of which have subsequently turned up in the real world.

For example, in his utopian book *New Atlantis*, published in 1627, the early British scientist Francis Bacon imagined a rational society guided by science, technology and trade, whose inhabitants benefited from refrigerators, airplanes and submarines. In his 1655 book *The States and Empires of the Moon and Sun*, the French writer Cyrano de Bergerac sent his hero to the Moon using a system of rockets lit in sequence, similar in concept to the multi-stage rockets used today.

It would take the dawn of the industrial revolution, however, for authors to start basing their futuristic speculations on the latest scientific knowledge. Jules Verne and H. G. Wells are the best-known nineteenth-century proponents of this approach, although neither would go so far as to let scientific reality get in the way of a good story.

In *From the Earth to the Moon*, Verne used a 300-m-long cannon to fire his characters to the Moon, even though he knew in reality that the sudden acceleration would kill them. In Wells's *The First Men in the Moon*, one of the characters invents a substance that cancels the effects of gravity and uses it to travel to the Moon.

Despite these occasional flights of fancy, both authors did accurately predict future technologies. Verne famously prefigured modern submarines in *Twenty Thousand Leagues under the Sea*, while Wells predicted nuclear energy and the atomic bomb in *The World Set Free*.

Since then, science-fiction authors have correctly predicted a whole range of future technologies, including solar power, robots, television, nanotechnology and waterbeds. In fact, so detailed was the US author Robert Heinlein's description of waterbeds in several of his science-fiction stories that the inventor of the first practical waterbed couldn't get a patent on it.

FROM SCIENCE FICTION TO FACT

In predicting the future, science-fiction writers have some important advantages over other technological forecasters. For a start, they come up with so much futuristic technology it's not surprising that some of it eventually becomes reality. They also don't have to specify a timescale for when the technology will appear. Another important advantage is that they can actually stimulate the development of their fictional technology.

Sometimes this is deliberate. In 1945, the British science-fiction writer Arthur C. Clarke published a serious proposal for a system of orbiting communication satellites. Although other scientists were thinking along the same lines, this proposal foreshadowed the current blanket of commercial and military satellites around the Earth.

Russian rocket pioneer Konstantin Tsiolkovsky's novel *Beyond the Planet Earth*, published in 1920, outlined many of the technologies subsequently adopted for space flight, including rocket fuel comprising a mixture of liquid hydrogen and oxygen. His ideas were influential on many future space scientists, especially in the Soviet Union.

But often it's not deliberate, it's just that many practising scientists and engineers are science-fiction fans and are inspired by what they have read about or seen (see Spotlight). The television programme *Star Trek* has been particularly influential in this respect: not only did it apparently give rise to automatic sliding doors, but *Star Trek* communicators clearly inspired the design of some early flip-up mobile phones.

Spotlight: Seeing the future

Science-fiction films have been around for almost as long as cinema itself, with French filmmaker Georges Méliès's *A Trip to the Moon* from 1902 one of the earliest examples. So, it's not surprising that the far futures envisaged by many of these films have since come to pass, allowing us to assess how accurate their predictions were.

As might be expected, most were way out. In 1997, Manhattan hadn't become a walled-in maximum security prison, as in *Escape from New York* from 1981, and in 2017 the US hadn't become a totalitarian police state showing game shows where convicted criminals battle to the death, as in *The Running Man* from 1987.

Others, such as *Back to the Future Part II* from 1989, were not quite so way out. While we didn't have the predicted flying cars or hoverboards in 2015, its vision of a society in thrall to modern technologies such as voice activation, display screens and fingerprint identification does seem eerily familiar. Adding to this familiarity is the fact that much of this future technology was attributed to companies that are still with us, such as Toyota, Nike and Pepsi.

Putting well-known companies into films set in the future is a common way for filmmakers to make their futures seem more realistic and plausible, but it is not without risks. This is demonstrated by two of the most famous science-fiction films ever made – *2001: A Space Odyssey* from 1968 and *Blade Runner* from 1982. Some of the companies that were included in these two films had gone out of business, become part of other companies or otherwise almost entirely disappeared from view by the time the future envisaged in these films had come to pass – 2001 and 2019 respectively.

This was the case for Pan Am, which supposedly operated the space shuttle in *2001: A Space Odyssey* and appeared on an electronic advertising hoarding in *Blade Runner*. Other examples include Howard Johnson's in *2001: A Space Odyssey*, Atari in *Blade Runner* and the Bell Telephone Company in both of them. Rather cleverly, when *Blade Runner 2049*, the sequel to *Blade Runner*, came out in 2017 it still referenced Pan Am and Atari, as a way to link to the original film and to emphasize that both films take place in an alternate future.

Key ideas

- Many widely predicted technologies either appeared much later than expected, such as videophones, or never at all, such as flying cars.
- Economics is perhaps the single most important factor in whether or not a new technology becomes a success.
- As well as many predicted technologies never arriving, other important technologies can arrive with no warning at all.
- The Second World War provided the impetus for the development of a whole host of unpredicted technologies, including computers and rockets.
- Science-fiction books, films and television programmes can sometimes stimulate the development of their fictional technology.

Dig deeper

Maynard, Andrew, *Films from the Future: The technology and morality of sci-fi movies* (Miami: Mango Publishing, 2018).

Tetlock, Philip, and Gardner, Dan, *Superforecasting: The art and science of prediction* (London: Random House Books, 2016).

27

A.I.

In science-fiction films, robots and intelligent machines tend to get a bit of a bad press. Either they're going mad and trying to kill us, like the computer HAL in the film *2001: A Space Odyssey;* or they've been programmed to kill us as part of some nefarious plot, like the android Ash in the film *Alien;* or they're trying to kill all of humanity because they think we're inferior, as happens in *The Terminator* films. If we're lucky, though, they may merely enslave us, as in *The Matrix* films.

There are obviously less hostile exceptions, like R2-D2, C-3PO and BB-8 in the *Star Wars* films and Data in *Star Trek: The Next Generation.* But even machines that are meant to serve humanity can be pretty deadly when required, such as the 'replicants' in the film *Blade Runner,* and so aren't to be trusted. Science fiction tells us that as soon as machines become more intelligent and powerful than humans, they quickly turn on us.

Despite these stark warnings, scientists and engineers are hard at work developing ever more sophisticated computers and robots, which can act, think and learn for themselves. This branch of computer science is known as artificial intelligence (A.I.).

Robots and intelligent machines

The aim is to produce machines that are just as intelligent, autonomous and dextrous as their fictional counterparts, but hopefully without the hostility. Such machines could help us do everything from solving currently intractable problems to dusting the furniture. Eventually, we may even develop machines that are, for all intents and purposes, conscious, although this would obviously raise a whole host of moral and ethical questions.

Spotlight: The spotter's guide to robots

In science-fiction films and books, a variety of terms are used to refer to autonomous machines: from robots to androids to cyborgs. So what is the difference between them?

Robots are essentially machines that move without direct human control; androids are robots that look like humans; and cyborgs are machines with both mechanical and biological parts. So R2-D2 is a robot; Ash and Bishop from the *Alien* films are androids; and the original Terminator is a cyborg.

THE LIMITATIONS OF SILICON CHIPS

To create such futuristic machines, scientists are not only producing more powerful versions of existing computers, but developing entirely new computing technologies.

As we saw in Chapter 18, modern computers are built around the semiconducting properties of silicon (and germanium), in the form of the ubiquitous silicon chip. By squeezing ever greater numbers of transistors onto silicon chips, scientists and engineers have produced increasingly powerful computers, able to conduct ever more complex operations in less and less time. But scientists are now starting to come up against physical limitations that will make it difficult for this to continue.

Indeed, even though transistors are still shrinking, the speed of silicon chips has levelled off since around 2005. The reason for this is that the speed of a silicon chip depends on how fast the transistors can be switched on and off, which depends on the size of the voltage applied to the transistor. We have now reached a point where this voltage cannot get much smaller and still reliably switch transistors on and off.

What is more, transistors may not shrink for much longer. Current methods for fabricating transistors, which involve carving them into silicon chips using beams of light, appear to be reaching their fundamental limits. Even if these limits can be overcome, as transistors shrink down to the molecular level, they will increasingly be exposed to the weird and wonderful world of quantum effects.

One of these quantum effects is electron tunnelling, whereby the probabilistic nature of electrons causes them to tunnel through physical barriers. As a consequence, electric current will start to leak across transistors even when they're supposedly switched off. So, before scientists have got anywhere near making a highly intelligent, conscious computer, silicon may be nearing its retirement.

New types of computer chips

Scientists are therefore working on new types of computer chips that would be less prone to these kinds of limitations. For example, they are looking at replacing silicon with nanomaterials like carbon nanotubes and graphene, and replacing electric currents with beams of light, which are faster (as they travel at the speed of light) but harder to manipulate and control.

Others are exploring even more radical approaches. If silicon transistors have problems operating at molecular scales, perhaps the answer is to replace them with actual molecules. This is the idea behind molecular computing and scientists have already produced molecular systems that can replicate various different logic gates, including AND, OR and NOT (see Chapter 18).

The basic approach is to design a molecule that only performs a certain action, such as conducting electricity or emitting light (output), if it binds with one or more other molecules (inputs). As with silicon-based logic gates, these molecular logic gates can be joined together to conduct more complex operations, such as addition and subtraction.

Logic gates have also been created from biological molecules such as DNA and RNA, which have the added advantage that they can act as a form of data storage. Here, the inputs tend to be biological molecules like enzymes and the output is a DNA or RNA strand that is modified in some way, such as being cleaved in two.

PARALLEL PROCESSING

But molecules don't just offer a new way to conduct conventional computing; they also offer the possibility of a totally new kind of computing. Current silicon-based computers can conduct simple operations very quickly, but only one at a time. This is mitigated by the fact that modern computer processors are compartmentalized, so that different sections can conduct different operations at the same time. Nevertheless, conventional computers are inherently serial: each section needs to complete an operation before it moves on to the next.

This means that computers are very good at performing complex mathematical calculations, but are less good at solving other kinds of problems, especially those that require parallel processing.

A good example is the 'travelling salesman' problem, which involves determining the shortest route between several cities. Courier firms face this kind of problem every day, but conventional computers can only solve it by painstakingly comparing every possible option. If there are a lot of cities, then even powerful computers struggle to come up with a solution.

On the other hand, a molecular computer could be designed to consider all possible options at the same time. This can be done by creating molecules or DNA strands that represent each solution and then getting them to react together in such a way that at the end of the process only the optimal solution is left. It sounds too good to be true, but molecular computers working in this way have already solved simple logical problems that silicon computers find tricky.

PROCESSING WITH QUBITS

For truly parallel processing, however, scientists are turning to the quantum world and replacing bits with qubits. As we saw in Chapter 18, a single bit can be either 1 or 0. By taking advantage of the fact that a quantum entity can exist in a 'superposition' of states (see Chapter 25), a single qubit can be both 1 and 0.

The implications of this are staggering: 10 bits can exist in over 1,000 different states (2^{10}), but only one at a time, whereas 10 qubits can exist in over 1,000 different states at the same time. Just 300 qubits could simultaneously exist in more states than there are atoms in the universe (2^{300} or 10^{90}).

Taking advantage of this massive parallel processing ability will be difficult, but by no means impossible. Scientists have already investigated various particles as possible qubits, including photons, electrons and individual ions. All you need is for the particles to possess a property that can exist in two states, representing 1 and 0; electrons, for example, have a property known as spin, which can be either up or down.

Scientists also have the technology to manipulate individual qubits, allowing them to be set up to conduct a calculation. For instance, individual atoms and ions can be manipulated by lasers and instruments known optical frequency combs, which produce light at specific frequencies.

USING QUANTUM CALCULATIONS

A more challenging problem is how to read out the answer of any quantum calculation, because a string of qubits can only exist in a superposition if it is not disturbed in any way. As soon as a scientist probes these qubits, the superposition collapses and each qubit adopts a fixed value of either 0 or 1, just like a bit. In a similar manner to molecular computing, this problem can be overcome by setting up the conditions in such a way that the superposition automatically collapses to the string of 0s and 1s that represent the correct answer.

A related problem is that any disturbance to the qubits, even being hit by a stray photon, will cause the superposition to collapse to a random mix of 0s and 1s. One solution to this problem is for a qubit to be represented by a collection of particles that all have the same quantum state, rather than by an individual particle.

Such collections can contain trillions and trillions of individual atoms. Because they all possess the same quantum state, they can still act as a single qubit and because there are trillions of them, they are much more resistant to disturbances. For example, a technique known as nuclear magnetic resonance can be used to put all the hydrogen nuclei (otherwise known as protons) in a flask of water into the same quantum state.

Using these kinds of techniques, companies such as Google and IBM have developed several generations of quantum computers with ever greater numbers of qubits, up to 72 at the time of writing. These early quantum computers have already proved adept at factoring numbers (breaking a number down to a string of smaller numbers that produce the original number when multiplied together). This is something that silicon-based computers find difficult to do for large numbers, which is why factoring forms the basis for the encryption systems that allow us to conduct secure financial transactions over the internet.

Researchers are now investigating various other potential applications for which quantum computers would be ideally suited. These tend to involve searching a large space of possible solutions, such as for predicting new materials and calculating financial risk.

Computers that are able to learn

Saying all this, scientists looking to develop intelligent machines are not quite yet ready to abandon silicon. Current computers may actually be more than complex and powerful enough to demonstrate high levels of intelligence, creativity and learning; we just need to find an effective way to program them with these kinds of abilities.

Already, scientists have produced computers that are able to learn. One way to achieve this so-called machine learning is via artificial neural networks, which aim to replicate the structure of the human brain. They consist of numerous nodes with set values connected together in a huge network that transforms inputs into specific outputs. Because the values of the nodes and the strength of the connections between them can be altered, the network changes as a result of experience, allowing the host computer to learn.

These neural networks are already being used to create computer systems that can learn to perform increasingly complex tasks, from diagnosing diseases to driving cars (see Spotlight). Such an approach could even one day lead to conscious machines. One of the possible reasons why humans are conscious while even the most powerful supercomputers aren't is our immersion from birth in a rich sensory environment that we can interact with. If machines had this kind of rich sensory experience, they too might naturally develop consciousness.

At which point, perhaps, we should become seriously concerned for our welfare.

Spotlight: Curiosity taught the computer

Rather ironically, machine learning currently relies quite a bit on human involvement. This is because in order for a computer system to learn how to play chess, identify a tumour or drive a car, it first needs to be trained.

This training is essential for setting the values of the nodes and the strengths of the connections in the system's artificial neural network, which is where the learning takes place. It either involves exposing the system to a load of inputs until it learns the desired outputs, such as accurately distinguishing between male and female faces. Or setting some specific objective within a controlled environment, such as winning a game, and then letting the system try out numerous different approaches for achieving that objective. In either case, humans are required to supervise and guide the training, providing the feedback that tells the system how well it is doing.

This approach has been very successful over the past few years, leading to the rise of virtual assistants like Amazon's Alexa and Apple's Siri and systems able to beat the best human masters of the Chinese game Go, but it requires a lot of human time and effort.

Now, though, computer scientists may have found a way to do away with the human trainers, by giving computer systems a very human quality: curiosity. This curiosity takes the form of deviations in predictions made by the system about data it is exposed to; the system is designed to find any deviations interesting and to explore them further. Such curiosity-driven systems have already shown great potential for learning how to play simple computer games without any instruction, because they quickly learn what many a gamer already knows: that dying in a game and going back to the start is boring.

Key ideas

- The branch of computer science looking to develop computers and robots that can act, think and learn for themselves is known as artificial intelligence (A.I.).
- Scientists are not only producing more powerful versions of existing computers, utilizing materials like graphene, but developing entirely new computing technologies, like molecular computing.
- The qubits used in quantum computing can exist in a 'superposition' of states, allowing a single qubit to be both 1 and 0 at the same time.
- Companies such as Google and IBM have developed several generations of quantum computers with ever greater numbers of qubits.
- Machine learning is achieved with artificial neural networks, which consist of numerous nodes with set values connected together in a huge network that transforms inputs into specific outputs.

Dig deeper

Fry, Hannah, *Hello World: How to be human in the age of the machine* (London: Doubleday, 2018).

Gribbin, John, *Computing with Quantum Cats: From Alan Turing to teleportation* (London: Black Swan, 2015).

28

Innerspace

Even if we manage to avoid being hunted to extinction or enslaved by super-intelligent computers, we could still be wiped out by swarms of mindless, nanoscale robots, or nanobots. This is the 'grey goo' scenario, as first postulated by the US nanotechnology pioneer Eric Drexler and then popularized in various works of fiction, including Michael Crichton's novel *Prey*.

The concern is that autonomous, self-replicating nanobots developed for some benign purpose, such as clearing up an oil spill, start to replicate out of control, consuming everything in their path. According to Drexler, if these nanobots replicate every 1,000 seconds, then within ten hours they would number almost 70 billion, within a day they would weigh 1 tonne and within two days they would weigh more than the Earth.

Now this is obviously rather far-fetched, especially as organic versions of these kinds of autonomous, self-replicating nanobots already exist, So far, however, bacteria and viruses have not yet consumed the entire planet in an orgy of replication, although they are pretty ubiquitous.

However, less destructive, non-reproducing versions of these nanobots are a more realistic prospect, especially if used in confined spaces such as within our bodies. Here, they would constantly monitor our health, looking out for signs of disease, cellular damage or things running out of control, like the abnormal cell division that leads to the growth of cancerous tumours. Working with the body's immune system, they would stop these processes in their tracks, banishing sickness and greatly extending our lives.

Treating cancer with gold nanoparticles

Although tiny robot surgeons and doctors are still a long way off, much simpler versions, designed to treat specific diseases and disorders, are already being developed. For example, scientists are taking advantage of the impressive abilities of nanoparticles (see Chapter 19) to find and treat cancerous tumours.

One way to do this is to get gold nanoparticles to congregate around a tumour and then shine infrared light at them; infrared light is used because it is better at passing through human tissue than visible light (otherwise we would be translucent). Absorbing this infrared light causes the gold nanoparticles to fluoresce at other infrared frequencies, allowing doctors to identify the precise location and extent of the tumour.

But that's not all, because absorbing infrared light also causes the gold nanoparticles to heat up, until they become hot enough to kill the cancer cells. A quick blast with infrared light produces an image of the tumour and a longer blast cooks it.

Obviously, the trick is to get the gold nanoparticles to congregate only around the tumour, thereby ensuring no damage to healthy cells. There are two ways this can be done. The easiest way is to take advantage of the fact that the blood vessels feeding tumours tend to have larger pores in their walls than normal blood vessels. Thus, all you need do is create gold nanoparticles that are too large to escape from normal blood vessels but small enough to escape from the blood vessels around tumours. A more targeted approach is to coat the gold nanoparticles with antibodies to proteins only found on the surface of cancer cells, or found in much greater concentrations.

Other nanoparticles can kill cancer cells in other ways, such as nanoparticles that contain iron, or bind with iron already in the body. This is because iron catalyses the production of oxygen-containing molecules known as reactive oxygen species. As their name suggests, reactive oxygen species are highly reactive, allowing iron-containing nanoparticles to act as tiny grenades that blast cancer cells apart.

Alternatively, the nanoparticles can carry anti-cancer drugs inside them or on their surfaces, releasing them when they arrive at a tumour. This offers a much safer way to deliver anti-cancer drugs, many of which are highly toxic to both healthy and cancer cells; hence, the intense physical suffering experienced by cancer patients undergoing chemotherapy. Early versions of this approach, in which the anti-cancer drug is enclosed within fatty bubbles called liposomes or natural blood proteins, are already being employed in clinics.

Using nanoparticles in this way allows tumours to be destroyed with far smaller quantities of these highly toxic drugs and without exposing healthy cells. It could even allow the use of drugs that are simply too toxic for use in conventional cancer treatments.

Again, the nanoparticles can be directed to the tumour by taking advantage of 'leaky' blood vessels or by attaching cancer-cell-targeting antibodies. Still, that leaves a number of other challenges to be overcome and scientists are exploring a whole range of options for doing that.

GETTING PAST THE BODY'S OWN IMMUNE SYSTEM

For a start, it's by no means certain that the nanoparticles will actually get to the tumour. Ironically, the reason for this is the body's own immune system, which as we saw in Chapter 9 is pretty quick to jump on any foreign particles. So, to reach the tumour, the nanoparticles first have to avoid being engulfed by macrophages or attacked by enzymes.

One way to do this is to house the nanoparticles within a material that the human immune system doesn't consider to be a threat and doesn't attack. This can obviously include biological material such as proteins and lipids, but it can also include certain polymers such as silicone.

Another way is to give the nanoparticles a means of propulsion, so they reach the tumour before the immune system has a chance to react to them. This can be done by utilizing some biomolecule naturally present in the body as fuel, such as glucose, or with a power source outside the body, such as a magnetic field. Another option is to attach the nanoparticles to a biological entity that can propel itself through the body, such as a bacterium with whipping flagella or sperm.

But assuming that the nanoparticle evades the immune system and reaches the tumour, how does it then release its anti-cancer payload? If the nanoparticle is permeable, then the drug could simply diffuse out, but that's no good if you only want to release the drug at the tumour. Or you could design the nanoparticles to degrade after a set period of time, ensuring this is long enough for the nanoparticles to reach the tumour, but that's a bit haphazard.

Alternatively, you design a nanoparticle that actively releases the drug only in response to a specific cue. This could be provided externally by a doctor; for instance, by shining infrared light at the nanoparticles. Or it could be provided internally by the environment around the tumour cells, which differs in certain ways from the environment around healthy cells, such as by being more acidic.

Spotlight: Enlisting the immune system

The immune system may be very good at attacking foreign invaders, whether viruses or nanoparticles, but it's much less good at attacking cancer cells. Scientists have known for a long time that cancer cells are able to evade the immune system, probably because they're mutated versions of our own cells rather than foreign invaders. Only recently, however, have scientists begun to discover the mechanisms that allow cancer cells to be so evasive, which has led to the development of a whole new approach to treating cancer called immunotherapy.

The immune system is a powerful weapon that can causes serious problems if it goes awry and attacks the wrong targets, as happens in autoimmune diseases such as multiple sclerosis. To try to prevent this happening, the system has various checks and balances. These include proteins on the surface of cells that can prevent the immune system launching an attack, even in the presence of antigens that should normally induce one. Like spies with forged papers, cancer cells tend to sport these 'do not attack' surface proteins, allowing them to allay suspicion.

Recently, however, scientists have begun developing drugs that can deactivate some of these surface proteins, exposing the cancer cells' subterfuge and giving the immune system full rein to launch an attack. Several of these drugs, known as checkpoint inhibitors, are now on the market and many more are in development. They work best on forms of cancer with lots of mutations, such as melanomas and lung cancer, as these cancers produce lots of antigens for the immune system to target.

They are also particularly effective in combination, or when used with conventional forms of cancer treatment such as chemotherapy and radiotherapy. What is more, they have inspired scientists to investigate other ways to enlist the immune system in the fight against cancer, including by developing vaccines based on cancer antigens and genetically modifying immune system cells.

Smart implants for other diseases

For other long-term diseases and disorders, the challenge is not so much releasing a drug at a specific time and place, but repeatedly releasing a drug when it's needed. To do this, scientists are developing the next generation of smart implants, able to activate appropriate treatments in response to some disease-specific cue.

For example, scientists are looking to develop implantable electronic devices that release life-saving drugs at the first physiological signs of an oncoming heart attack, which could be abnormal behaviour of the heart or the presence of certain characteristic proteins in the blood. If these devices were made from flexible semiconducting plastics, they would be able to mould to the interior of the body.

The behaviour of the heart is already being monitored by the latest implanted pacemakers and defibrillators, both of which stimulate the heart with electricity to prevent it beating abnormally. These pacemakers and defibrillators regularly communicate wirelessly with a base station about both their activity and the heart's behaviour. The base station then sends this information over the internet to a doctor, allowing early intervention if the heart starts to fail.

Eventually, our bodies could contain whole suites of clever implants, constantly monitoring our vital functions, wirelessly communicating with each other and the outside world, and automatically triggering the appropriate treatment, from electrical stimulation to the release of drugs. This system will probably then evolve into the suite of constantly circulating medical nanobots mentioned earlier.

Stimulating tissue growth

Until that time, however, we will continue to suffer from disease and disorders, some of which can result in the loss of large amounts of tissue and even whole organs. At the moment, the only option is to replace the lost tissue and organs with transplanted material, either from the patient's own body or from others, usually after death. But obviously the amount of spare material that can be harvested from a person's own body is limited and the supply of donated organs is inherently unreliable.

So, scientists are working on ways to grow spare tissues and organs in the laboratory. For instance, scientists are now able to stimulate the growth of excess bone on a patient's normal bone; this excess bone can then be harvested and used to replace bone lost as a result of disease or accidents.

An alternative approach is to implant some kind of scaffold into a patient's body to give the tissue cells something to grow on. This is commonly done to fix small holes in the heart: a fabric is placed over the hole and heart cells then grow over the fabric, which eventually degrades. But scientists are now working on doing this at much larger scales and in three dimensions.

MATERIALS FOR GROWING NEW CELLS

They are doing this with a variety of materials, including hydrogels, shape-memory polymers and electrospun nanofibres. Hydrogels are highly absorbent gels consisting of an interlinked network of long molecules, either synthetic or biological, in which water can account for up to 99 per cent of the weight, giving them a consistency similar to biological tissue. Shape-memory polymers can be squashed to make them easy to insert into a body, before naturally reverting to their natural shape in response to some environmental cue, usually the temperature of the body. Electrospun nanofibres are pulled out of a polymer solution with an electric force, just like candyfloss.

The idea is to impregnate these materials with cells, often some form of stem cell, and growth-promoting chemicals, so that they provide a three-dimensional scaffold for the growth of new cells. The advantage of having lots of different materials as potential scaffolds is that different cell types like to grow on materials with different properties, and often respond in different ways to different materials, allowing the scaffold to be tailored to the cell and the desired behaviour.

Scientists have successfully grown biological tissue such as skin, neurons, bone and cartilage on these materials. Eventually, they would like to grow – or even print with 3D printers – whole organs in the laboratory, using the patient's own cells, and then implant these organs into the patient. They are already part of the way there: 3D printers can deposit mixtures of cells and polymers that grow into living tissue, while simple versions of organs, known as organoids, can be produced using induced pluripotent stem cells (see Chapter 20).

Enhancing our bodies

But perhaps all this new implant technology could do more than simply keep us healthy, perhaps it could actually improve us. The fictional counterpart here is the 1970s television programme *The Six Million Dollar Man*, in which a former astronaut is given bionic limbs and eyes that provide him with enhanced strength, speed and vision.

Scientists are now developing such bionic limbs, but are currently finding it difficult enough to replicate our existing physical abilities. Nevertheless, certain technologies being developed would give us enhanced abilities, such as contact lenses with built-in displays. In addition, scientists have developed versions of biological material that are tougher than the real thing, such as bone grown on carbon nanotubes. There is even talk of implanting electronic devices into the brain to enhance intelligence and memory.

Improving our bodies in this way may be essential if we are to undertake the next great technological leap: travel into the depths of space.

Key ideas

- Scientists are taking advantage of the impressive abilities of nanoparticles to find and treat cancerous tumours.
- Nanoparticles can be directed to a tumour by taking advantage of 'leaky' blood vessels or by attaching cancer-cell-targeting antibodies.
- Immunotherapy enlists the immune system in the fight against cancer, by blocking the surface proteins that cancer cells use to evade it.
- Scientists are developing the next generation of smart implants, able to activate appropriate treatments in response to some disease-specific cue.
- Scientists are working on ways to grow, or even print, spare tissues and organs in the laboratory for implanting into patients.

Dig deeper

Graeber, Charles, *The Breakthrough: Immunotherapy and the race to cure cancer* (London: Scribe Publications, 2018).

Piore, Adam, *The Body Builders: Inside the science of the engineered human* (New York: HarperCollins, 2017).

29

A space odyssey

In films, traversing the vast expanse of interstellar space presents no problem at all. You just leap into the USS *Enterprise* or *Millennium Falcon*, turn on your warp drive or make the jump into hyperspace and, hey presto, you've soon arrived at another star system. In reality, not only are we obviously far from developing this kind of spacecraft technology, but it may actually be impossible for us to do so.

That does not mean, however, that humankind will not at some point travel across the whole galaxy. It just may take a bit of time.

Even to colonize the galaxy at a leisurely pace, we'll need to develop novel kinds of spacecraft and propulsion systems. At the moment, getting into space requires rockets, which work on the same principle as a firework. A chemical propellant is transformed into a hot gas, comprising lots of fast-moving molecules that are directed out the back of the rocket, moving it in the opposite direction.

Getting a rocket into space

This movement is all down to the principle of action and reaction, as first formulated in the seventeenth century by the British scientific genius Isaac Newton. According to Newton's third law of motion, for every action there is always an equal, opposing reaction. This means that if you shoot hot gas out the back of a rocket, it moves forward.

The hot gas doesn't need to press against anything to provide this movement; if it did, then rockets wouldn't be able to move in the emptiness of space. Simply applying a force in one direction produces an equal force in the opposite direction.

Numerous different kinds of propellant can be used in rockets, but the one employed in many space-bound rockets is a mixture of liquid hydrogen and liquid oxygen. Reacting liquid hydrogen and oxygen together at high temperatures and pressures produces a hot blast of water vapour, which is directed out the back of the rocket at speeds of up to 4.5 kilometres a second (km/s).

Keeping a rocket travelling in space

This is sufficient to accelerate a spacecraft to the speed needed to escape from the Earth's gravitational pull, which is 11.2 km/s (over 40,000 kilometres an hour). But once in space, the rockets generally stop firing, often detaching and falling back to Earth, and the spacecraft simply coasts along at the same speed. It doesn't slow down because it is travelling through an almost empty vacuum, meaning no friction. While most current spacecraft do have their own small rocket engines, these are mainly used for manoeuvring.

A spacecraft can still accelerate once in space, however, by performing gravitational slingshots, in which it picks up speed by skirting around a planet. In this way, the spacecraft essentially captures some of the planet's momentum as it orbits the Sun. Performing several gravitational slingshots around Venus and Earth allowed NASA's *Cassini* spacecraft to travel to Jupiter at a cruising speed of almost 32 km/s.

TRAVELLING BETWEEN THE PLANETS

Even at these speeds, travelling between the planets takes a long time. It may only have taken three days for astronauts to travel the 384,400 km to the Moon, but it takes at least 180 days for current spacecraft to travel to Mars. Travelling the 4.3 light years

(40 trillion kilometres) to our nearest star, Proxima Centauri, at these speeds would take 40,000 years.

These journeys would be much quicker if a spacecraft could continue firing its rocket engines once it entered space. This would allow it to accelerate for longer and thus reach higher speeds, but simply requires too much propellant. To accelerate a single atom to a speed of 10,000 km/s (3 per cent the speed of light), taking us to Proxima Centauri in 130 years, using current chemical rockets would require 10^{434} atoms of propellant, which are many more atoms than exist in the observable universe.

Getting there faster

FISSION ROCKETS

So, if we want faster travel to the planets in our solar system, and especially if we ever want to travel beyond our solar system, then we need other means of propulsion. One alternative is to replace chemical rockets with nuclear rockets. As we saw in Chapter 24, nuclear reactions are a million times more efficient at generating energy than chemical reactions.

The simplest approach would be a variation of existing rockets. A fission reactor would be used to heat hydrogen to a temperature of several thousand degrees, producing fast moving hydrogen molecules that are fired out the back of the rocket. Tests on prototype fission rockets performed in the 1960s and 1970s indicated that hot hydrogen molecules could be expelled at speeds of up to 70 km/s. Even faster ejection speeds (perhaps up to several thousand kilometres a second) could theoretically be generated by replacing the fission reactor with a fusion reactor, but as we discovered in Chapter 21 scientists are still unable to produce a sustainable fusion reaction.

ION PROPULSION

Various practical problems, together with the high expense, explain why US scientists abandoned the development of fission rockets in the 1970s. But the recent successful deployment of a new kind of propulsion system, known as ion propulsion, could open the door to a new kind of nuclear-powered spacecraft.

In ion propulsion, an electrical generator strips electrons from the atoms of a gas such as xenon, transforming the atoms into positively charged cations. By attracting them towards a negative electrode, these cations are fired out the back of the spacecraft, providing thrust.

Even though the cations can be expelled at speeds of up to 100 km/s, the thrust they generate is minuscule: equal to keeping a piece of paper in the air by blowing on it. Current ion propulsion engines take four days to accelerate a small spacecraft from a standstill to 100 kilometres an hour. This means that spacecraft with ion propulsion engines still need to be launched into space using conventional rockets.

The reason for the lack of thrust is that although the cations are expelled at great speed, there aren't that many of them. But that means not only can the amount of xenon gas used as the propellant be quite small – less than 30 kg would take a spacecraft to Mars – but the engines can be fired all the time, providing constant – if gentle – acceleration. Over time, this gentle acceleration can produce impressive speed increases.

NASA's *Dawn* spacecraft used an ion propulsion engine to travel to the asteroid belt, visiting the giant asteroid Vesta in 2011 and the dwarf planet Ceres in 2015. This engine increased the spacecraft's speed to more than 10 km/s over the course of its eight-year mission and allowed *Dawn* to become the first spacecraft to orbit two planetary bodies.

Replacing the electrical generator with a fission reactor could theoretically increase ejection speeds to 1,000 km/s, sufficient to take a spacecraft to Proxima Centauri in 100 years. But this would require huge amounts of uranium to fuel the fission reactor: a 10-tonne spacecraft (about four times the size of *Cassini*) would require 50 million tonnes of uranium, many times greater than the Earth's known uranium resources.

The depressing reality is that in order to reach our nearest star in less than a human lifetime, we need a spacecraft that could reach speeds of at least 30,000 km/s (10 per cent the speed of light) and this is well beyond our current technological capabilities. That has not stopped scientists thinking up possible ways to achieve this goal, however.

USING ANTIMATTER TO GENERATE ENERGY

One idea is to take advantage of antimatter, which is exactly the same as normal matter but with opposite physical properties. So, a positron is exactly the same as an electron, but with a positive rather than a negative charge. When a particle meets its antimatter equivalent, they annihilate each other, transforming all their mass into energy in accordance with Albert Einstein's famous equation $E = mc^2$ (see Chapter 16).

The annihilation of antimatter thus represents the most efficient way possible to generate energy, being thousands of times more efficient than nuclear fission. If antimatter annihilation is used to convert a liquid propellant into a hot gas, then accelerating a one tonne spaceship to 30,000 km/s would require only four tonnes of propellant and around 12 kg of antimatter.

Unfortunately, we are far from being able to produce even that small amount of antimatter. At the moment, antimatter is only produced in particle accelerators, at a rate of just a billionth of a gram a year.

THE SOLAR SAIL AND THE RAMJET

If the perennial problem is the amount of propellant, perhaps the best option is to do away with the propellant. This is the reasoning behind both the solar sail and the ramjet.

A solar sail takes advantage of the fact that photons of light exert a small force when they bounce off a reflective surface, and so consists of a thin, reflective surface tethered to a spacecraft. The idea is that photons emitted by the Sun reflect off the solar sail, exerting a combined force large enough to generate forward momentum, just like wind on a conventional sail.

Obviously, the solar sail would need to be massive – several hundred metres across – to reflect sufficient photons to accelerate itself and an attached spacecraft. But it does conjure up an attractive retro-future image of astronauts majestically sailing around the solar system. By also taking advantage of the Sun's gravity, these astro-sailors could use the solar sail to travel in any direction, even towards the Sun. Although this sounds like science fiction, the first prototype solar sail, just 20 m across diagonally, was deployed in space in June 2010.

A solar sail could also theoretically take a spacecraft to another star, but in this case the sail would need to be 30 km wide and it would need to be driven by a more intense source of photons than can be provided by the Sun. The idea would be to build a giant solar-powered laser in space. This would fire an immensely powerful laser beam at the solar sail, accelerating it to 60,000 km/s (20 per cent the speed of light) over the course of 30 years, allowing it to reach Proxima Centauri in 50 years.

With an even larger, multi-component solar sail (around the size of France), the spacecraft could even be slowed down as it reached its destination and then returned to Earth. If a central section of the solar sail popped out behind the main sail and light was reflected from the main sail onto this smaller sail, then the force of the laser would slow the craft down. This larger sail could also accelerate the spacecraft to 50 per cent the speed of light (150,000 km/s).

In essence, a ramjet would simply be a fusion-powered rocket, but one that collected all its fuel from space. As we saw in Chapter 1, hydrogen is by far the most common element in the universe, forming dense molecular clouds and also widely dispersed throughout interstellar space.

By extending a magnetic field as big as the Earth in front of it, a ramjet would funnel all the available hydrogen into its fusion reactor, using the energy generated to expel the fusion products out the back. The really clever bit is that as the ramjet speeds up it collects more hydrogen, allowing it to accelerate even more. The faster it goes, the faster it will be able to go. Calculations show that after a year of travelling, a ramjet could accelerate to 270,000 km/s (90 per cent the speed of light).

As the ramjet accelerates, it will get closer and closer to the speed of light, but it will never be able to exceed it. Because, as Albert Einstein proved, nothing can travel faster than the speed of light: 300,000 km/s is the universe's ultimate speed limit. This means that the idea of hopping in your spaceship and quickly zooming over to another star system will probably remain a fiction (although see Spotlight).

Spotlight: Taking the shortcut

We may not be able to travel faster than the speed of light if taking the usual route through the universe, but what about a shortcut?

Wormholes could possibly provide just such a shortcut. They are breaks in the fabric of space joining two regions that might be separated by thousands of light years. It's as if you're standing in a first-floor bedroom and want to get to the kitchen directly beneath you. You could walk down the stairs to the kitchen, but it would be faster if you could jump through a hole in the bedroom floor.

Although wormholes have never been detected, their existence does not contravene our current physical theories. Unfortunately, if they do exist, they're probably highly unstable, only lasting for a fraction of a second. Scientists have suggested, however, that something called 'exotic matter', which quantum theory suggests is constantly being created and annihilated at tiny scales, could stabilize wormholes, potentially allowing a spacecraft to pass through.

GENERATION SHIPS

Even at light speed, a trip to Proxima Centauri will take over four years, while getting to the centre of our galaxy, the Milky Way, would take 27,000 years (although see Spotlight). But that doesn't mean that humankind won't eventually colonize the galaxy, even at much slower speeds. Using massive spaceships containing hundreds or thousands of people and entire ecosystems, multiple generations could travel between the stars, even if a single journey took thousands of years. This is the idea behind so-called generation ships.

Spotlight: Not so fast

Although it would take a spacecraft travelling at the speed of light 27,000 years to reach the centre of our galaxy, it wouldn't seem that long to those on board. This is because, as Albert Einstein again proved, the faster you move, the more time slows down.

This ensures that nothing can travel faster than the speed of light. Otherwise, if you were travelling in a spacecraft at half the speed of light and turned on the headlights, the subsequent beam of light would be travelling at one and a half times the speed of light. But it doesn't, because time slows down.

At low speeds, this slowing is barely perceptible, although it has been experimentally proven by flying highly accurate atomic clocks in airplanes. At close to the speed of light, though, the effects become dramatic.

At 86 per cent of the speed of light, time elapses at half its normal rate and at 99.5 per cent of the speed of light it elapses at only one-tenth its normal rate. Indeed, to those on board a spaceship travelling close to the speed of light, it would appear to take just 20 years to reach the centre of the galaxy.

Interstellar travellers could travel to a star system and colonize it. A thousand years later, some of their descendants could then start the journey to a new star system. In this way, it has been estimated that even if it took several thousand years to travel between stars, humankind would spread over the whole galaxy in 100 million years. This may sound like a long time, but the dinosaurs bestrode the Earth for longer.

Spotlight: No space for us

A space-faring future for humankind will not just require novel propulsion systems, but also novel ways to keep humans safe and healthy during their space travels. Humans evolved on a world with gravity, oxygen, liquid water, food and protection from cosmic rays. Space has none of those things, and that will pose an increasing problem as humans spend more time in space.

Without gravity to work against, astronauts' bodies lose both bone density, up to 2 per cent a month, and muscle bulk, up to 5 per cent a week. The lack of gravity also alters the distribution of water and blood in the body. Not only can this give astronauts a puffy appearance, but it can also put more pressure on the backs of their eyeballs and compress their optic nerve, potentially putting their vision at risk over long timescales.

For some reason, zero-gravity can cause some normally harmless bacteria and microbes inhabiting the human body to become pathogenic, which has led to astronauts developing respiratory and dental infections. This is made worse by the fact that astronauts' immune systems are often weakened by exposure to the cosmic rays that suffuse space (see Chapter 2). Comprising energetic particles such as protons and atomic nuclei produced by the Sun and other stars, these rays can also generate reactive oxygen species in the body, which can go on to damage blood vessels, cause cancer and promote further bone loss.

There are ways to lessen these effects, but they can't be eradicated completely. Regular exercise helps to slow the rate of bone and muscle loss, while drugs that soak up reactive oxygen species might help to protect against cosmic rays. More dramatic solutions, inspired by science fiction, could be required, such as spinning part of a spacecraft to simulate gravity.

Key ideas

- Rockets work by transforming a chemical propellant into a hot gas that is fired out the back of the rocket, moving it in the opposite direction.
- In ion propulsion, an electrical generator strips electrons from the atoms of a gas such as xenon, transforming the atoms into positively charged cations that are fired out the back of the spacecraft.
- In order to reach our nearest star in less than a human lifetime, we need a spacecraft that could reach speeds of at least 30,000 km/s.
- A solar sail takes advantage of the fact that photons of light exert a small force when they bounce off a reflective surface.
- As Albert Einstein proved, nothing can travel faster than the speed of light: 300,000 km/s is the universe's ultimate speed limit.

Dig deeper

Collins Peterson, Carolyn, *Space Exploration: Past, present, future* (Stroud: Amberley Publishing, 2017).

Wall, Michael, *Out There: A scientific guide to alien life, antimatter, and human space travel (for the cosmically curious)* (New York: Grand Central Publishing, 2018).

30

Things to come

'In the past century, there were more changes than in the previous thousand years. The new century will see changes that will dwarf those of the last.' So said the British science-fiction writer H. G. Wells (who also wrote the screenplay of the 1936 film that gives this chapter its title) in a lecture he gave in 1902.

His prediction was correct and will probably continue to be correct for the foreseeable future, as our rate of technological development keeps on increasing. The nineteenth century brought railways, electricity, photography and the first motor cars, but the twentieth century brought airplanes, nuclear power, genetic modification and the internet. Imagine what we'll achieve in the next 100 years or the next 1,000 years or the next 10,000 years. Providing, that is, we don't destroy ourselves in the meantime (see Chapter 24).

Imagining what we'll achieve in the future is not just the domain of science-fiction writers. Several scientists have also made predictions for how humankind might progress and of the advanced technologies we might develop. Unfortunately, the spectre of that other major British science-fiction writer, Arthur C. Clarke, hangs over these predictions. Specifically, his famous dictum: 'Any sufficiently advanced technology is indistinguishable from magic.'

The temptation when making these predictions is to assume that within a few thousand years humankind will be so technologically advanced as to be able to do almost anything, no matter how inconceivable it seems at the moment. Unlike many science-fiction writers, scientists try to resist this temptation by ensuring their predictions don't contradict any of our current physical theories. Their predictions may be practically impossible with our current technology, but they shouldn't be theoretically impossible.

That's not to say, however, that in a few thousand years our science and technology won't have developed to a stage where we are able to achieve the currently inconceivable, such as travel faster than the speed of light. But when making predictions, it's often useful to have a few self-imposed restrictions. So, based on our current understanding of science, all the advanced technologies detailed in this chapter are theoretically possible.

Colonizing space

What almost everyone looking into the future of humankind agrees on, whether science-fiction writers or scientists, is that at some point we'll make a major push into space, probably using some of the novel propulsion technologies detailed in Chapter 29. Now this may be a risky prediction to make, seeing as in the 1960s many believed that we would have bases on the Moon and perhaps even Mars by the beginning of the twenty-first century. At some point, however, the advantages of colonizing space will start to outweigh the disadvantages.

THE ASTEROID BELT

Although it may take a few hundred years, we will eventually reach the point when humankind starts to need more living space or natural resources than can be provided by the Earth. The simplest way to resolve both of these shortages may be to turn to the large rocks making up the asteroid belt, which orbits between Mars and Jupiter.

Because these asteroids formed from the same disc of dust and gas as the Earth, they are rich in many of the same materials as found on Earth, including iron, silicon, nickel, copper, lead and zinc, as well as carbon, water, hydrogen and nitrogen. So, when we've used up all the deposits of these materials that are easily extractable from the Earth's crust, we could start obtaining them from asteroids.

This would either involve establishing mining colonies on the asteroids while they're still in the belt, several thousand of which are known to have diameters greater than 1 km, or transporting the asteroids back to the Earth and dismantling them in orbit. This transporting could be achieved by simply firing bits of the asteroid into space. According to the principle of action and reaction (see Chapter 29), this would accelerate the remaining bulk of the asteroid in the opposite direction, potentially taking it all the way to Earth.

Furthermore, in an extreme form of recycling, after an asteroid was hollowed out by mining it could subsequently be used as a habitat. The idea would be for colonists to live on the inside walls of the asteroid, which would be set spinning to simulate the force of gravity. Taken together, the thousands of asteroids in the belt with diameters greater than 1 km have a surface area greater than that of all the continents on Earth.

SPACE HABITATS

Alternatively, the resources derived from asteroids could be used to construct tailor-made space habitats. Scientists have already come up with potential designs for such habitats. In the 1970s, a US physicist called Gerard O'Neil designed a cylindrical space habitat, in which people would live in three equally spaced strips running down the length of the cylinder. These habitable strips would be separated by three equally sized strips of transparent wall.

On the outside, three aluminium mirrors would run the length of the cylinder, reflecting sunlight through the transparent walls onto the habitable strips. In addition, these mirrors could be moved to cover the transparent walls, blocking off all incoming light and simulating night. As with the asteroids, the habitat would spin about its axis, with one full rotation every minute, to simulate gravity, while all its energy requirements would be supplied by an array of solar panels deployed at one end of the cylinder.

O'Neil envisaged these space habitats coming in several different sizes. The smallest would be 1 km long and 200 m wide, housing 10,000 people; the largest would be 32 km long and 6.4 km wide, housing 1 million people. Attaching engines to these habitats would turn them into the kind of star-travelling generation ships mentioned in Chapter 29.

LIFE ON MARS

If we became really short of land, either because the human population had grown too large or because we had made large areas of the Earth uninhabitable, then our only option would be to move to Mars. Before we did this, however, we would have to make Mars habitable, because it currently lacks liquid water or a breathable atmosphere and has an average surface temperature of –65°C. Such terraforming, as it is known, does seem to be feasible, although it would be a massive undertaking.

According to a study conducted by NASA scientists in the early 1990s, tens of billions of tonnes of a greenhouse gas such as the ozone-destroying chlorofluorocarbons (CFCs) would need to be pumped into the thin atmosphere of Mars. This would thicken up the atmosphere and instigate global warming, increasing the planet's average temperature by around 20°C. As a consequence, both the polar ice caps and the permafrost that covers the surface of Mars would start to melt, releasing trapped carbon dioxide and thereby enhancing global warming. More greenhouse gases could be supplied by seeding the planet with genetically modified or synthetic bacteria designed to thrive in these harsh conditions.

Focusing solar energy onto the polar ice caps would transform the melting ice into water vapour, further enhancing global warming. This water vapour would condense to form clouds and then torrential rain, eventually producing rivers, seas and oceans on the surface of Mars. Finally, plants would be introduced to release oxygen into the atmosphere. In total, this terraforming effort would probably take over 1,000 years.

Capturing more of the Sun's energy

By then, it won't just be land and resources that humankind could be running short of, but also energy. All our current energy sources are ultimately derived from the Sun, but the Earth receives less than one billionth of the energy produced by the Sun. In the future, we should be able to capture more of the Sun's energy by constructing huge solar panels in space, but eventually that would not be enough.

Between 1973 and 2007, humankind's consumption of energy grew by an average of just over 0.01 per cent a year. If that rate of increase continues, within 300,000 years we will be using the equivalent of all the energy released by the Sun.

To capture all of the Sun's energy, our only option would be to build a huge sphere around it, with a radius equal to the distance from the Sun to the Earth. Humankind could then live on the inner surface of this sphere, which would have room for hundreds of billions or even trillions of people. This is known as a Dyson sphere, after the US physicist Freeman Dyson, who first came up with the idea.

In actual fact, a solid sphere of that size would not be stable, so Dyson suggested a system of rings that together would fully encircle the Sun. Obtaining the material needed to construct these giant rings would unfortunately require taking apart several of the planets in the solar system.

Spotlight: Olaf Stapledon, 1886–1950

Of all the science-fiction writers that have attempted to peer into the far future, perhaps the most ambitious was the British writer Olaf Stapledon. In fact, Stapledon was as much a philosopher as a science-fiction writer, awarded a doctorate in philosophy from the University of Liverpool in 1925 and writing several academic works on philosophy and ethics.

As a consequence, he approached the far future from a unique perspective, and he regarded his two novels about the far future – *Last and First Men* and *Star Maker* – as 'future histories' rather than conventional science fiction. He was interested in producing a speculative but plausible account of the future prospects for humankind and the universe, rather than simply writing stories set in the future.

What marks out *Last and First Men*, published in 1930, and *Star Maker*, published in 1937, is both their scope and foresight. *Last and First Men* describes the future of humankind over 2 billion years, during which time numerous civilizations rise and fall, humankind evolves through 18 different species that differ greatly in their appearance and our home world shifts from the Earth to Venus to Neptune. But *Last and First Men* is far eclipsed by *Star Maker*, which follows the development of life in the universe over 100 billion years, and describes many different alien civilizations and cultures.

In these novels, Stapledon also introduced many concepts that subsequently became staples of both science fiction and mainstream science, several of which are described in this chapter. So, *Last and First Men* contains the first description of terraforming, when the fifth species of men make Venus habitable, and an early mention of genetic engineering, when the eighth species transform themselves into a form that can live on Neptune. *Star Maker* contains a version of what came to be known as Dyson spheres, with Freeman Dyson freely admitting that the novel was a key inspiration.

The death of the Sun

In the very long term, however, even Dyson spheres won't be enough to safeguard the future of humanity. For a start, in about 6 billion years' time, our Sun will expire, turning first into a huge red giant and then into a tiny spent star known as a white dwarf. Hopefully by then, humankind will have colonized the whole galaxy, making the death of the Sun and the destruction of our home planet less of a trauma than it otherwise might have been. Humankind may well have also split into numerous different species, driven by both natural evolution and genetic engineering, with each species adapted to the conditions in their particular star system.

So long as there are stars in the galaxy, humankind will probably have a plentiful supply of energy. Unfortunately, however, that won't always be the case. As we learned in Chapter 16, entropy within the universe is steadily increasing: usable energy is being transformed into unusable, diffuse heat. In practical terms, this means that the huge molecular dust clouds that are the birthing grounds of new stars are gradually being used up. According to calculations, the era of star formation will come to an end in around one hundred trillion years.

THE END OF THE UNIVERSE AS WE KNOW IT

At that point, the Milky Way will consist of white dwarfs, brown dwarfs (small stars that are unable to ignite nuclear fusion in their cores), planets, asteroids and massive black holes (see Chapter 2) swallowing up everything else. For quite a long time, around 10^{27} years, we could conceivably derive energy from these massive black holes, but they too will eventually evaporate away.

It's not even as though we could escape to other galaxies. Not only because the same processes would probably be occurring there, but because those other galaxies would by then be too far away. Recent astronomical observations have indicated that the expansion of the universe is actually speeding up, perhaps due to the action of 'dark energy' (see Chapter 1). If that is the case, then at some point in the distant future all the other galaxies that we now see would have disappeared over the horizon (apart from our nearest galaxy neighbour Andromeda, which may well collide with the Milky Way in around 6 billion years). We would be utterly alone in the universe.

So is there no way for humankind to survive as the universe gradually winds down? Well, there might be. Some scientists theorize that when a massive star collapses to produce a black hole it may trigger the formation of a new universe, completely separate from our existing universe. If humankind could somehow travel through a black hole, perhaps one that we have created, then it might end up at the very dawn of a new universe.

And that, I believe, is where we came in.

Key ideas

- When we've used up all the deposits of materials such iron and silicon from the Earth's crust, we could start obtaining them from asteroids.
- Humans could live in hollowed-out asteroids or in tailor-made space habitats constructed from asteroid material.
- Terraforming Mars to make it habitable for humans would probably take over 1,000 years.
- To capture all of the Sun's energy, humans could build a huge sphere around it, known as a Dyson sphere, and then live on the inner surface.
- Due to entropy, the era of star formation should come to an end in around one hundred trillion years.

Dig deeper

Galfard, Christophe, *The Universe in Your Hand: A journey through space, time and beyond* (London: Pan Books, 2016).

Kaku, Michio, *The Future of Humanity: Terraforming Mars, interstellar travel, immortality, and our destiny beyond* (London: Penguin, 2019).

Index

* Page numbers in italics refer to illustrations